Modern Bacterial Taxonomy

Modern Bacterial Taxonomy

Second edition

Fergus Priest
and
Brian Austin

Department of Biological Sciences
Heriot-Watt University, Edinburgh, UK

CHAPMAN & HALL

London · Glasgow · New York · Tokyo · Melbourne · Madras

Published by Chapman & Hall, 2–6 Boundary Row, London SE1 8HN

Chapman & Hall, 2–6 Boundary Row, London SE1 8HN, UK

Blackie Academic & Professional, Wester Cleddens Road, Bishopbriggs. Glasgow G64 2NZ, UK

Chapman & Hall Inc., One Penn Plaza, 41st Floor, New York, NY10119, USA

Chapman & Hall Japan, Thomson Publishing Japan, Hirakawacho Nemoto Building, 6F. 1-7-11 Hirakawa-cho. Chiyoda-ku, Tokyo 102, Japan

Chapman & Hall Australia, Thomas Nelson Australia, 102 Dodds Street, South Melbourne, Victoria 3205, Australia

Chapman & Hall India. R. Seshadri. 32 Second Main Road. CIT East. Madras 600 035, India

First edition 1986 Van Nostrand Reinhold (UK) Ltd

Second edition 1993

© 1986, 1993 Fergus Priest and Brian Austin

Typeset in 10/12 Palatino by Fox Design, Bramley, Surrey
Printed in Great Britain by the Alden Press

ISBN 0 412 46120 X

A catalogue record for this book is available from the British Library

Library of Congress Cataloging-in-Publication data available

∞ Printed on permanent acid-free paper, manufactured in accordance with the proposed ANSI/NISO Z 39.48-199X and ANSI Z 39. 48-1984

Contents

Preface to the second edition

Much has happened in the seven years since the last edition of this book. Numerical taxonomy and traditional chemotaxonomy have been eclipsed by developments in molecular biology which have led microbial systematics into the new era of molecular systematics. The rate at which nucleic acid sequencing and related techniques proceed is remarkable, and the applications which emerge are equally impressive. This development has warranted a complete new chapter to describe techniques such as the polymerase chain reaction and its applications, use of restriction fragment length polymorphisms in various guises, and the progress in RNA sequencing and its uses. These developments are having far-reaching implications in our understanding of such diverse topics as the early evolution of life on this planet, the basic phylogenetic lineages of all living organisms and the ecology of non-culturable micro-organisms which can only be detected and enumerated using nucleic acid hybridization probes.

Needless to say, in true taxonomic fashion these developments have not been without controversy. Heated debates continue on the 'best' way to classify micro-organisms, should we use phylogeny or phenetics or try to combine both approaches? Should the archaebacteria be assigned to their own 'domain' of equal status with the true bacteria and the entire eukaryotic world including animals, plants and protists? Answers to these issues are emerging, but it is unlikely that consensus will be achieved in the near future. We believe that this is a healthy attitude because it stimulates debate in a topic which is perceived by many as dull, and drives research forward into new fields.

In the second edition of this book, we have updated all the chapters and introduced a new chapter on nucleic acid analyses in microbial systematics. We have attempted, where appropriate, to emphasize the benefits of the 'new microbial systematics'. We have tried to show how new techniques in microbial classification have either given us new insight into the relationships of micro-organisms and their evolution or have been used to improve bacterial identification. We have also expanded considerably the chapter dealing with applications of systematics in other disciplines such as ecology, pathogenicity and biotechnology. It is largely through emphasizing the importance of bacterial systematics to bacteriology in its entirety that the topic can be taught to students in a stimulating fashion, and we hope that the second edition of this book will help instructors provide such courses and thus foster a more widespread appreciation of the importance of bacterial systematics and the exciting, vital questions now being addressed in the subject.

We are grateful to numerous colleagues for the provision of material for inclusion in this book, in particular Dr Martin Maiden and Professor Alan Yousten and to the staff of Chapman & Hall. Finally, our gratitude to Professor Michael Goodfellow, who taught us both microbial systematics as undergraduate students and without whose enthusiasm and encouragement this book was unlikely to have been written.

Fergus Priest and Brian Austin

Introduction

There has long been a fascination with the relationships among living things and their arrangement into categories, beginning with the writings of Aristotle, developed by the great naturalists of the last century, and advancing now into molecular evolution through the application of molecular biological techniques.

The early formulation of relationships was accepted without question, largely because living organisms could be readily distinguished by morphology. It may seem obvious that a dog is distinct from a horse, and although the reasons may be difficult to explain, they will include assessment of size and shape differences. In the case of bacteria, the problem is exaggerated because of the extremely small size of the organisms and lack of pronounced morphological variation. It was indeed astonishing that van Leeuwenhoek, the founding father of microbiology, managed to observe what are now regarded as bacteria-like objects, considering that the maximum magnification of his simple microscope did not exceed 300 times. The 'animalcules', recorded in his landmark publication of 1684, constituted the first legitimate report of microbes (bacteria?) and he made careful illustrations that, to some extent, suggested morphological variations between the cells. Nevertheless, it took many more years for classifications of bacteria to be developed. Perhaps some of the delay stemmed from a genuine lack of interest in taxonomy. Much of the early microbiology was conducted by medical scientists; Pasteur, for example, was more interested in the physiology and pathogenicity of micro-organisms than in their relationships to each other; an attitude which prevails to this day.

Early classifications were based on largely morphological information, as determined by light microscopy. Indeed, the different morphological forms of bacteria were described lucidly by Müller in the eighteenth century. Thereafter, credit must be given to Ehrenberg and Dujardin for their seemingly revolutionary classifications. For example in the 1830s before the advent of pure culture techniques, Ehrenberg described the family 'Vibrionia' and divided it into four genera.

Following the interest in morphology, pure culture techniques allowed more rigorous study of bacterial metabolism, and physiological information was included in classifications. A classic example is the work of Orla Jensen, who in 1919 divided the lactic acid bacteria into four genera which, with modifications, are used today. Thus *Lactobacillus*, *Streptococcus*, *'Betacoccus'* (now *Leuconostoc*) and *'Tetracoccus'* (now *Pediococcus*) were classified on the basis of physiological (e.g. type of fermentation, growth temperatures) and morphological criteria. A particularly readable account of these early contributions to microbiology has been published by T. D. Brock (*Milestones in Microbiology*, American Society for Microbiology, Washington, DC, 1975).

More recently, the trend has been to supplement the physiological information with data from chemical analyses and molecular biology. In particular, informational macromolecules such as DNA, RNA and proteins act as repositories of evolutionary development. Chromosomes accumulate mutations and diverge from common ancestors with time. This is of course reflected in the sequences of stable RNA molecules (transfer (t) RNA and ribosomal (r) RNA) and proteins. Comparisons of gene sequences allow the reconstruction of evolutionary pathways; the less similar two equivalent sequences, the longer ago they diverged from the common ancestor and began accumulating independent mutational differences. This macromolecular phylogeny has led to some startling discoveries such as the archaebacteria which are believed to be direct descendants of the most ancient living organisms, and is revolutionizing our ideas about the early stages of evolution on this planet.

This book provides an introduction to bacterial classifications with particular emphasis on modern trends in the subject and development of molecular systematics. It also covers the more traditional approaches to the subject such as the use of numerical techniques to quantify relationships based on morphology, physiology and chemical structure. Most importantly, it includes recent approaches to the identification of bacteria based on nucleic acid hybridization probes.

At this stage, it will avoid confusion to define some key terms. Taxonomy (often considered to be synonymous with classification) is regarded as the theory of 'classification' (this concerns the arranging of organisms into groups); 'taxon' (pl. taxa) is a group of individuals of any rank. Genera, families and species are taxa. Nomenclature is the process of allocating names to the taxa. Identification is the means by which unknown organisms are allocated to previously described taxa. All these terms apply to systematics, which is the study of the diversity and relationships among organisms.

CHAPTER 1

Classification

What are the purposes of classifications? Why do we bother to arrange bacteria or higher organisms into groups? There are, in fact, at least three important answers to this question. The first concerns the transfer of information. Quite simply, a classification is a means of summarizing and cataloguing information about an organism. It follows that a classification is a form of database or information retrieval system containing a large amount of information about an organism which is collected and summarized by its place in the scheme. Its position in the system is denoted by the use of a name. For example, the generic name *Bacillus* indicates a group of Gram-positive bacteria which differentiate into endospores under aerobic conditions. More specifically, *Bacillus subtilis* refers to a group of strains with the general characteristics of the genus, which in addition secrete several extracellular enzymes such as amylases and proteases, use nitrate as a terminal electron acceptor under anaerobic conditions and are naturally competent for DNA-mediated transformation. This ability to predict properties of groups is a very important aspect of classifications, particularly in these days of computerized databases in which large amounts of information can be stored and easily accessed. Indeed, the more information in the classification the more useful it is.

A second purpose for classification is that organisms must be categorized into groups before identification systems can be devised for the recognition of new isolates. It is evident that without prior arrangements of individuals into groups it would be impossible to assign new isolates to a taxon, and without such order, we would not be able to conduct science.

The third purpose of classification may be considered to provide an insight into the origins and evolutionary pathways of bacteria and higher organisms. Indeed for some people, evolution is of over-riding importance and the evolutionary pathways and the classification are considered one and the same; but this need not necessarily be the case.

If classifications are to serve these purposes effectively, they should fulfil several criteria. Firstly, a classification should have high information content: essentially the greater the amount of information on which it is based, then the greater will be the 'predictivity' of the

1

classification and the more generalizations that may be made about the taxa involved. Secondly, classifications should be stable. This may seem obvious, but a classification, in which the composition and descriptions of taxa change frequently is confusing and unhelpful. Finally, it is important that the development of a classification should be empirical, reproducible and scientifically based.

1.1   DEFICIENCIES OF TRADITIONAL CLASSIFICATIONS

It is apparent to students of microbiology that bacterial classifications have, until recently, failed to achieve the necessary requirements of predictivity, stability and objectivity. One reason for this is that these aspects of a classification are intimately entwined and failure to achieve one is generally reflected in failure to satisfy all three. It is pertinent to enquire precisely what the early bacterial taxonomists were doing wrong so that we may be able to improve the situation. These shortcomings have been fully discussed by Cain (1962), but need to be briefly reviewed here because it is important to understand the reasons for the failure of early classifications.

Classifications have been traditionally based on Linnaeus's principles, which suggested that the process of classification should be conducted from 'above' and, by starting from the overall groups encompassing all living things, repeated divisions could be made until the species level was reached. In this system, species were recognized as being indivisible, i.e. the basic taxonomic unit, and at every rank, taxa were defined by specific features, which reflected the 'essential nature' of the groups. In the case of bacteria, these features might be the presence or arrangement of flagella, the morphology of the aerial mycelium in *Streptomyces* or the ability of *Escherichia coli* to form acid from lactose. The taxonomist had simply to discover these features and to distinguish the important from the unimportant to effect classification. This is regarded as an *a priori* choice of characters, since it supposes that the important feature(s) of a group can be chosen deductively. However, the process is purely subjective, since it cannot be known by intuition which features best reflect the 'essential nature' of the group or, for that matter, the organism. When such 'important' features are discovered, the reason is usually that the group has been already subjected to systematic study and useful diagnostic characters have been highlighted. The *a priori* choice of characters leads to serious problems because of disagreement between scientists. Characters considered by one worker to be of inestimable importance for defining groups may be totally disregarded by others. Thus, a lengthy discourse occurred earlier this century concerning the relative importance of morphological characters, such as flagella pattern, cell shape, and physiological characters for classification. Indeed, this approach was taken to the extreme in the misguided assumption that a progression of characters existed that defined the hierarchy of taxa. It was considered that morphological characters defined genera, physiological characters defined species and serological features could be used to define sub-species.

If the criteria of a good classification as discussed above are considered, it will be apparent that the traditional approach failed in all respects. Since only a few so-called 'important' characters were used to construct the classification, it was based on little information and lacked predictivity. Very few assertions could be made about the taxa. Also, the classifications were unstable because the choice of important characters was subjective. Different taxonomists expressed contrasting views about the composition and defining features of taxa. This resulted in the continual revision of taxa with new descriptions and quite often new names. Finally, it may be argued that classification has not been conducted as an empirical science, because of reliance on subjectivity and intuition in the choice of defining characters. It was not repeatable because of the involvement of personal judgements by the scientists.

Having argued against the traditional approach to classification, it is necessary to provide a satisfactory alternative, but to precede this, consideration must be given to the kinds of classification that are available.

1.2 THE RANGE OF CLASSIFICATIONS

It is important to emphasize at this point that there is no single unifying classification of living organisms. Biologists are dogmatic in their belief that there is a single correct way to classify the individuals of a population. Sneath (1983) suggested that 'biology is so complex that underlying explanations are very difficult to detect: therefore there is a demand for a single dominating concept that will encompass all of its phenomena'. Alternatively, a view has been expressed that life may have started as a single event (whether caused by Creation or natural process) and consequently there has been a tendency to think of a single plan underlying all living organisms that can be used as a basis for classification. However, there is no reason to require a single classification; classifications are devised by man for various purposes. Since we have many purposes in mind there are many types of classification, and it is possible to classify them! Essentially there are three types of classification dependent on the nature of the relationships used in their construction.

1.2.1 Special-purpose classifications

Bacteriologists often use classification and identification schemes designed for their particular discipline. For example, food micro-biologists or insect pathologists might use specific identification systems for the bacteria that they are most likely to encounter. Justifiably, these schemes ignore all other bacteria as irrelevant. They may be extremely useful for the specialist microbiologist, but are of little value to microbiology in general because most bacteria are excluded.

Special-purpose classifications are artificial in that they seldom dis-play the 'natural' relationships among the organisms. The distinction between *Shigella dysenteriae* and *Escherichia coli* is a well known example. Strains of these taxa share considerable DNA sequence homology and are phenotypically very similar. From virtually every viewpoint they could be considered as a single species. However, the more serious pathogenicity of *S. dysenteriae* is of considerable importance to the clinician (and the patient!), and consequently, the separate taxa have been retained. Similarly, *Bacillus cereus* and *Bacillus thuringiensis* are virtually identical, but differ in one important respect. Strains of *B. thuringiensis* contain a large plasmid which carries a gene encoding a protein toxin which is lethal to various insect larvae. This bacterium is used to control insect pests of agricultural crops and therefore it is useful to indicate its single difference from *B. cereus* by using a differ-ent name. These artificial divisions obviously have their uses, and, in the two examples cited above, the needs of the specialist have been recognized by microbiologists in general, and these opinions are incorporated into the mainstream of microbiological classification. It is necessary for the clinical bacteriologist to distinguish *S. dysenteriae* from *E. coli* and for the insect pathologist to recognize the insect patho-genic bacilli; but we must also recognize the limitations of this approach. Artificial classifications are monothetic, in that a single feature (patho-genicity in the above examples) is deemed both sufficient and necessary for the placement of an organism in a group. Thus, for inclusion of a bacterium in the species *S. dysenteriae* it must cause dysentery in man. Non-pathogenic strains cannot be included in this taxon.

Monothetic classifications suffer from the serious disadvantage noted for traditional classifications in section 1.1. They are based on restricted information; pathogenicity in the cases mentioned above. They also tend to be unstable. Bacteriologists with different interests adopt different schemes, leading to considerable confusion. For example, the plant saprophyte *Erwinia herbicola* (classified and named by plant pathologists) is synonymous with an intestinal organism, *Enterobacter agglomerans* (classified and named by clinical microbiologists). After much debate about which is the 'correct' name, the latter won and the name *E. agglomerans* is used today. Finally, identification schemes derived from monothetic classifications readily lead to misidentification. Because the group is based on a single or few features, the unknown organism need only be aberrant in that feature to be assigned to the wrong taxon. Relatively non-pathogenic isolates of *S. dysenteriae* would be placed in the genus *Escherichia*, non-toxic, plasmid-deficient strains of *B. thuringiensis* will be identified as *B. cereus*. Artificial classifications have their uses, but as a general system of value to all microbiologists their limitations are severe.

1.2.2 Natural (phenetic) classifications

The alternative to the special-purpose classification is the general-purpose classification, a system that is of value to all microbiologists whatever their discipline. Such a classification should encompass all bacteria and all aspects of these bacteria. Since special-purpose classifications are artificial so general-purpose classifications can be described as 'natural'. Natural in this sense can be attributed to Gilmour (1951) and was developed by Sneath (1983) to refer to relationships between organisms based on their overall similarity or affinity. Natural relationships embody all aspects of the organisms from molecular structure through physiology, to habitat. Such relationships are termed phenetic and refer to affinities based on the complete organism (genotype and phenotype) as it exists at present with no reference to the evolutionary pathways or ancestry of the organism. This contrasts with the term natural used in its evolutionary context (section 1.2.3).

In phenetic classifications, organisms are arranged into groups (phena) on the basis of high overall similarity using both phenotypic and genotypic characters. This approach encompasses all measurable features of the organisms so that the resultant classification should be useful to all microbiologists. Moreover, the taxa are polythetic rather than monothetic because they are defined as having a high number of features in common and there is no requirement for the presence of a particular attribute. Thus, individuals aberrant in a specific character can be accommodated by such groups. In a phenetic classification, *E. coli* and *S. dysenteriae* would be placed in the same species, since they have high genotypic and phenotypic similarity. Consequently, the distinction attributed to relative pathogenicity would not be given undue importance. This describes the natural relationships of these bacteria, and, although it may alarm the clinical bacteriologist, it is a sensible approach insofar as strains of *E. coli* carrying toxin and surface antigen-encoding plasmids are associated with serious diarrhoeal disease. This fact is often obscured by the artificial division of these organisms into 'pathogenic' and 'nonpathogenic' types.

It follows that phenetic classifications do not suffer the shortcomings of their artificial counterparts. Because the classification is based on the overall properties of the organisms it has a high information content with the associated predictivity. The classification is also more stable, since the same names should be used to describe the same taxa regardless of the interests of the microbiologist. Finally, since phenetic classifications are always, in theory, polythetic, identification should be more accurate. Individuals are identified to groups using several characteristics and organisms that do not fully conform are still accommodated by such groups because identification requires only that the organism has the overall characteristics of the group. One favourite example is the black swan of Australia that is assigned to the swan species because it has the overall characters of the swan despite being the 'wrong' colour.

1.2.3 Natural (phylogenetic) classifications

For many biologists, particularly those who study animals and plants, the term 'natural' refers to an intrinsic quality of nature and natural classifications are based on 'species or groups of species that exist in nature as a result of a unique history of descent with modification' (i.e. evolution) (Wiley, 1981) or 'the true historical entities produced by the evolutionary process' (Cracraft, 1983). These natural classifications are based on phylogenetic (genealogical) relationships in that they attempt to trace the evolutionary pathways that have given rise to the organisms as we view them today, with the classification exactly reflecting the line of ancestry. This classification will be congruent with the phenetic classification if there has been no parallel or convergent evolution and the rate of change proceeds constantly in all lineages. The two classifications will differ, however, if convergent evolution or recent gene transfer gives rise to organisms that are phenetically similar but have different ancestry (Fig. 1.1). Much of what follows is more relevant to plant or animal biologists than microbiologists, but with the advent of macromolecule sequencing, bacteria are becoming more amenable to

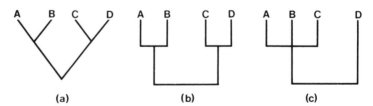

Fig. 1.1 Three dendrograms representing (a) a cladogram indicating the phylogenetic relationships of taxa A,B,C,D, (b) a phenogram of the same taxa in which evolution has been assumed to be divergent and at constant rate and (c) a phenogram in which convergent evolution or recent gene transfer has resulted in C being phenetically related to A,B.

phylogenetic analysis and therefore the major schools of phylogenetic thinking are described briefly below.

Among phylogeneticists, there is considerable controversy. The term cladistic refers to the branching pattern that describes the pathway of ancestry of a group of organisms. Hennig (1966) showed how cladistic relationships might be inferred. Hennigean cladistics involves the determination of monophyletic groups (or clades) that can be defined as groups distinguished by a set of characters inherited from an ancestor that are not shared by any other species outside that clade. In other words, all members of a monophyletic group possess a homologous character either in its primitive (earlier) or derived (later) form and it is the joint possession of this character by all descendants of the species that defines the group. In this choice of important characters, it can be seen that Hennigean cladistics differ completely from the pheneticist's view of a natural classification. As an example of Hennig's approach, Fig. 1.2 shows three organisms A, B and C classified on the basis of

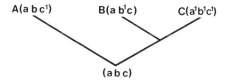

$A(a\ b\ c^1)$ $B(a\ b^1c)$ $C(a^1b^1c^1)$

$(a\ b\ c)$

Fig. 1.2 A cladogram of species A, B and C. B and C constitute a monophyletic group defined by the shared derived character b^1. A and B are not monophyletic because species C is excluded. Species A and C are not monophyletic because they have acquired character c^1 independently; it is not a homologous character (after Williams, 1985).

three characters a, b and c. B and C comprise a monophyletic group defined by the shared character b^1 derived from the primitive character b via an ancestor (not shown) at the upper branch point. b/b^1 is therefore a homologous character. A and B are not monophyletic because C, which does not share the primitive feature a, is excluded from the group. Moreover, A and C are not monophyletic because the only character which they have in common was acquired independently, i.e. is non-homologous. It is argued that by careful selection of characters, non-convergent homologous characters and monophyletic groups can be determined and built into a hierarchy that must reflect the evolutionary pathway. Strict adherence to this process, often termed evolutionary cladistics, can lead to some startling relationships. For example, birds and crocodiles form a natural clade because they share a common ancestor (Fig. 1.3). Crocodiles and lizards, on the other hand, do not form a clade because they share a common ancestor with the birds which is not included in the class Reptilia.

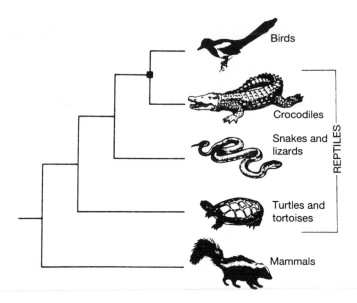

Fig. 1.3 Cladogram of birds, reptiles, and mammals. The reptiles do not constitute a natural clade because they share ancestors with the birds, which are not included in the Reptilia. Birds and crocodiles, on the other hand, constitute a natural clade (Archosauria) because they share a common ancestor (black box) not shared by any other organism (reprinted from Li and Graur (1991) with permission).

'Traditional evolutionists' avoid this problem by following the ideas of Simpson (1961) and Mayr and Ashlock (1991) and dilute phylogenetic relationships with an element of phenetic similarity when constructing the classification. Classification is practised with reference to the phylogeny, but without the requirement that all groups be monophyletic. The difference between Simpsonian and Hennigean cladistics has been neatly summarized by Williams (1985): 'According to Simpson (1961) the classification should be consistent with the presumed phylogenetic relationships, whereas for Hennig they are one and the same thing'.

To confuse the issue further, 'transformed cladists' or natural order systematists have drifted from Hennig's approach to such an extent that they are more akin to pheneticists and their views have no connection with phylogeny. They argue that evolution cannot be known and evolutionary homologous characters are unknown and can only be deduced from the classification itself. As emphasized by Sneath (1983) 'homologies cannot be recognized from the way character states are distributed in monophyletic groups because these groups cannot be constructed until the homologies have first been recognized'. The approach of the transformed cladists is therefore to assess the distribution of various character states among a group of organisms and to arrange the organisms into a classification determined by just one criterion, maximum parsimony, that is the route which involves the minimal number of changes to arrive at the simplest possible arrangement (see Chapter 5). Thus the 'natural order' of taxa is revealed without recourse to any evolutionary theories other than that of maximum parsimony. Such arguments are fortunately of little relevance to mainstream microbiology because traditional characters useful for the construction of phylogenetic groups are few. Moreover, it is not possible to determine those characters which are primitive and those which are derived and attempts at constructing phylogenies of bacteria on the basis of morphological or physiological characters have been uniformly disastrous for this reason (see Chapter 5). However, recent developments in molecular biology have enabled phylogenetic relationships to be deduced on the basis of macromolecular sequence analysis. This approach has revolutionized much of our thinking about classification of bacteria and indicated how cladistic classifications can be constructed (see Chapter 5). Cladistic classification has therefore become an issue in microbiology and again caused lengthy debate about the relative merits of this approach compared with phenetics.

1.3   MERITS OF PHENETIC VERSUS PHYLOGENETIC CLASSIFICATIONS

Since it is commonly (but incorrectly) upheld by taxonomists that there is only a single 'natural' classification of organisms, there is obviously considerable controversy among the different schools of thought as to which approach should be adopted. In the case of bacteria, should classification be phenetic, based on overall similarities of genotype and phenotype as we view them today, or should it be cladistic, based on the evolutionary pathways deduced from macromolecular sequences? It will be useful to consider the relative merits of phenetics and phylogenetics in this context.

1.3.1   'Goodness' of the classification

The basic criterion of the phylogenetic classification is that it should precisely reflect the evolutionary pathway of the organisms. However, it is impossible to compare the cladogram with the true cladogeny because the latter is unknown (merely inferred from the cladogram) and probably can not be known.

The goal of the pheneticist is less well defined, but it could be argued that the classification should represent, as accurately as possible, the affinities between each and every organism. Various statistical measures have been developed to test the distortion within a hierarchical classification and optimum procedures for constructing a phenetic classification have been proposed (see Chapter 2). However, because of the difficulties in defining the ultimate phenetic classification, we can not evaluate the accuracy of the classification. We must discuss the relative merits of phylogenetic and phenetic classifications on some other grounds.

1.3.2   Verifiability

Since the construction of a classification is a scientific exercise it should be a testable hypothesis or contain testable hypotheses. The phylogenetic approach is not verifiable in this sense because the only way to test a cladogram is with a second cladogram based on a different data set. As pointed out by Sneath (1983), if both cladograms have been derived using the same assumptions they may well be congruent, but this does not test those assumptions. It merely shows that the two databases are compatible and that there has been no lateral gene transfer but it remains impossible to compare the outcome with the actual evolutionary pathways. Moreover, all current theories of evolution may seem incorrect to the next generation of biologists.

Conversely, the phenetic classification, perhaps because its aims are less ambitious, is more accessible to verification. The operation of producing a phenetic classification involves gathering data, which are then analysed using established statistical techniques (see Chapter 2). This process is entirely objective and can be repeated (verified) by a second scientist. Moreover, phenetic classifications are independently testable. As new data are generated by advances in science so they can be included in the classification. If the original classification was correct, in that it accurately represented the affinities of the organisms, the new data should not alter it. If however, the new information does affect the classification it can be included and a composite, improved classification can be constructed. Phenetic classifications are therefore continually being updated and refined to produce the most accurate representation of the phenetic relationships under study.

1.3.3   Practicalities

It has been emphasized by Sneath (1989) that, if a phylogenetically coherent group was found to be phenetically heterogeneous then the cladistic group would be of little practical value. Conversely, if two phylogenetically independent lines converged due to strong selection pressure such that the organisms were phenetically similar, there would seem to be little practical value in separating them. Thus, the pheneticist aims for homogeneous taxa which can be readily identified. The argument of the cladist is that phylogenetic groups, once determined, often prove to be phenetically coherent on further study. Indeed, were this not so, phylogenetic classifications would be of little value outside the specialist field of genealogy.

One of the principal aims of a classification is to provide a scheme whereby unknown organisms may be identified. For the microbiologist, simple, reliable, rapid tests are needed. Phenetic classifications can be analysed to select the most diagnostic characters for delineation of groups and to provide reliable identification schemes (Chapter 7). Phylogenetic classifications of bacteria rely largely on gene sequence data. Recent developments in hybridization technology are offering simple identification procedures based on nucleic acid probes which can be extremely specific and accurate. Although there remains a requirement for reasonably sophisticated technology, developments in this area are certain to continue to simplify the procedures and this is certainly one of the most important practical applications to emerge from molecular systematics (Chapter 7).

1.4   THE CHOICE BETWEEN PHENETIC AND PHYLOGENETIC CLASSIFICATIONS

The requirement is for a stable classification of high information content (predictivity) that lends itself to practical applications (e.g. identification of new strains; databases) and from which phylogenetic relationships can be inferred. Despite various claims in the literature, there is no firm evidence that cladistic classifications are any more predictive, congruent or stable than phenetic classifications (Sokal, 1985), and, in view of the comments made previously, it is more likely that phenetic classifications are superior on these grounds. Indeed, by changing the methods of analysis of sequence data, far-reaching changes in the branching patterns of cladograms can ensue (see Chapter 6). The choice therefore must be primarily for phenetic classification. It is important to note that many of the current approaches to bacterial classification based on ribosomal (r) RNA sequences and related techniques purport to be phylogenetic, but are actually phenetic measures of affinity with molecular sequences as characters. Inferences can be made about evolutionary patterns from these studies, but supposed evolutionary pathways are not necessarily shown in the classification. For such studies to show cladistic patterns the analyses must be made using phylogenetically based algorithms.

We agree with Jensen (1983) who suggested that what is needed are:

1. classifications that reflect what is known about the taxa, and
2. procedures for generating hypotheses about evolutionary relationships.

To this end, many systematists now agree that the two systems should be combined as far as possible (Stackebrandt, 1988). A classification framework derived from cladistic analysis of macromolecular sequences is of value to those interested in evolutionary patterns but the Hennigean reference system (containing branching patterns alone) is of limited general use. Phenetic studies (including both phenotypic and genotypic information) provide the practical databases but offer little insight into evolution. By combining both approaches, the framework can be derived from phylogenetic relationships and the detailed information from phenetics. Studying bacteria, we are fortunate that the items to be classified, the species, are largely the same whether defined phenetically or phylogenetically (see Chapter 6). It is only the arrangements of these units that will differ according to the relationships used for the classification, and in many areas the phylogenetic and phenetic are congruent. In several instances, the congruity was not obvious, for example the transfer of the thermoactinomycetes from the actinomycetes to the family Bacillaceae on the basis of rRNA sequence analysis at first seemed illogical. What could these branching filamentous bacteria have in common with the rod-shaped endospore formers? However, the relatively low content of guanine and cytosine in the DNA, the peptidoglycan composition of the cell wall and the formation of true endospores by thermoactinomycetes showed that these bacteria had considerable phenetic homology with the thermophilic bacilli. Perhaps this phenetic relationship would have been discovered earlier had the pheneticists not been clouded in their outlook by the 'importance' of morphological features, such as cell shape.

Problems arise when phylogenetic and phenetic arrangements differ and in these cases, we suggest that the classification should favour the phenetic. We see little value in a purely cladistic classification that does not accommodate the degree of phenetic relationship between species. Such is the case with higher organisms as we saw in Fig. 1.3. There seems little point in emphasizing the cladistics of the reptilia to the exclusion of the phenetic viewpoint and indeed most would agree that the phenetic approach, which groups the reptiles together and excludes the birds, is appropriate for most uses.

We shall investigate both aspects of bacterial taxonomy in the ensuing chapters and explore ways in which they can be integrated.

A simplified classification of the bacteria has been included in Appendix A, and illustrates the enormous complexity of the subject.

REFERENCES

Cain, A.J. (1962). The evolution of taxonomic principles, *Symposium of the Society for General Microbiology*, **12**, 1–3.

Cracraft, J. (1983). The significance of phylogenetic classifications for systematic and evolutionary biology, in *Numerical Taxonomy* (J. Felsenstein, Ed.), pp. 1–17, Springer-Verlag, Berlin.

Gilmour, J.S.L. (1951). The development of taxonomic theory since 1851, *Nature*, **168**, 400–402.

Hennig, W. 1966. *Phylogenetic Systematics*. University of Illinois Press, Urbana.

Jensen, R.J. (1983). A practical view of numerical taxonomy or should I be a pheneticist or cladist?, in *Numerical Taxonomy* (J. Felsenstein, Ed.), pp. 53–71, Springer-Verlag, Berlin.

Li, W.H. and Graur, D. (1991). *Fundamentals of Molecular Evolution*. Sinauer Associates, Sunderland, MA.

Mayr, E. and Ashlock, A.D. (1991). *Principles of Systematic Zoology*, McGraw Hill, New York.

Simpson, G.G. (1961). *Principles of Animal Taxonomy*. Columbia University Press, New York.

Sneath, P.H.A. (1983). Philosophy and method in biological classification, in *Numerical Taxonomy* (J. Felsenstein, Ed.), pp. 22–37, Springer-Verlag, Berlin.

Sneath, P.H.A. (1989). Analysis and interpretation of sequence data for bacterial systematics, *Systematics and Applied Microbiology*, **12**, 15–31.

Sokal, R.R. (1985). The principles of numerical taxonomy 25 years later, in *Computer-Assisted Bacterial Systematics*, (M. Goodfellow, D. Jones and F.G. Priest, Eds.), pp. 1–20, Academic Press, London.

Stackebrandt, E. (1988). Phylogenetic relationships vs. phenotypic diversity: how to achieve a phylogenetic classification system of the eubacteria, *Canadian Journal of Microbiology*, **34**, 552–556.

Wiley, J.E.0. (1981). *Phylogenetics: The Theory and Practice of Phylogenetic Systematics*. Wiley, New York.

Williams, J. (1985). Cladistics and the evolution of proteins, in *Computer-Assisted Bacterial Systematics*, (M. Goodfellow, D. Jones and F.G. Priest, Eds.), pp. 61–90, Academic Press, London.

CHAPTER 2

Numerical taxonomy

2.1 INTRODUCTION

The early part of this century witnessed dramatic advances in biochemistry, genetics and other aspects of biology, but taxonomy plodded on in a haphazard way and contributed little to biology as a whole. Until the late 1950s, bacterial taxa were still recognized on the basis of a few 'key' features, and heterogeneous ill-defined genera were commonplace. Identification of unknown isolates was almost impossible, with the exception of a comparatively few medically important species. With the advent of the computer age, and thus the ability to manipulate large amounts of data rapidly, a development occurred that revolutionized the approach to bacterial taxonomy. Sneath introduced the concept of numerical taxonomy in 1957, and many hundreds of publications have since been devoted to the topic (see Sneath and Sokal, 1973).

Numerical taxonomy, also referred to as NT, Adansonian taxonomy (after the eighteenth century French botanist Michael Adanson), computer taxonomy, numerical phenetic analysis, taxometrics and taxonometrics, is defined by Sneath and Sokal (1973) as: 'the grouping by numerical methods of taxonomic units into taxa on the basis of their characteristics'. Following the basic principles of Adanson, numerical taxonomy necessitates studying as many aspects (traits or characters) of the biology of organisms (referred to as operational taxonomic units: OTUs) as possible. This generates a mass of information on the OTUs under study. However, a key feature of numerical taxonomic methods is that, *a priori*, all the characters have equal importance or weight, a concept that has been difficult for many conventional microbiologists to accept. Indeed this aspect of numerical taxonomy has undoubtedly led to some of the fiercest arguments. Classical taxonomists insist that some tests are more important than others for defining taxa, and that these should be used to establish the classification. Conversely, numerical taxonomists insist that all characters should have equal weight in the construction of classifications. Once the taxa are defined, however,

14

1 **STRAIN SELECTION**
 - Need for pure cultures
 - Inclusion of replicates
 - Inclusion of reference cultures

2 **TEST SELECTION**
 - Variety of tests
 - Minimal number about 50
 - Adoption of rapid methods where possible

3 **RECORDING RESULTS**
 - Analysis of test error and rejection of poorly reproducible tests

4 **DATA CODING**

5 **COMPUTER ANALYSES**
 - Calculation of similarities
 - Cluster analysis

6 **INTERPRETATION OF RESULTS**
 - Definition of clusters
 - Identification scheme
 - Selection of representative strains for allied studies

Fig. 2.1 Stages in numerical taxonomy analysis of bacteria. One of the advantages of numerical taxonomy is that it emphasizes this logical progression of steps involved in classification.

useful differential characters may be weighted (*a posteriori*) for identification purposes.

It is an important premise that taxa should be defined on the basis of overall similarity, and not created simply as a result of taxonomic intuition. This means that taxa are not established solely by the presence (or absence) of certain pre-determined 'essential' characters among groups of OTUs. Thus individual characters, such as the presence of yellow pigmented colonies, are not sufficient to justify the inclusion of OTUs within taxa and the groups are defined polythetically.

Figure 2.1 is a flow diagram of the stages involved in numerical taxonomy. These stages include strain and test selection, coding of data and their entry into the computer, data analyses, and interpretation of results. The various stages will be considered separately.

2.2   STRAIN SELECTION

With the availability of comprehensive software packages for numerical taxonomy (Sackin, 1987), the procedures may be applied to large numbers of bacterial strains. It is often possible to cope with upward of 600 OTUs in a single set of analyses. Larger data sets may have to be divided, but overall comparisons can still be made by means of inter- and intra-group analyses. In other circumstances, the total number of OTUs may be quite small, but as a general guideline, data sets should contain information on at least 60 strains. Smaller numbers of OTUs have been used, but apart from studies of very specialized groups or for demonstration purposes, it is questionable whether computer-based numerical taxonomy procedures were justified in such cases.

Essentially, studies may be categorized as those that are 'broadly' or 'narrowly' based (restricted studies). The former seek to study large groups of organisms with a view to defining taxa; usually species and genera. These may comprise named strains, such as representatives of bacterial families, e.g. Enterobacteriaceae, or unnamed isolates such as those recovered from ecological studies. Of course, great care is needed in the initial selection of isolates otherwise the outcome of the investigation may be meaningless. These sets of OTUs should include some strains of known identity for comparative purposes. The reference strains serve as markers, and ease the ultimate identification of unknown organisms. It is advantageous to include at least two reference strains of each species, as this reduces the possibility of mistakes caused by contamination or mis-labelling, which would negate their comparative value. Such numerical taxonomy studies have been used successfully to examine representative isolates from a diverse array of natural habitats, including soil, leaf surfaces and water, and to rationalize large heterogeneous taxa, such as *Streptomyces*. Examples of narrowly based or restricted studies include the investigation of taxonomic relationships and validity within groups of related organisms, for example *Staphylococcus*. These may explore intra- and inter-generic relationships. In most of these cases, named cultures will be used; and it is recommended that the *bona fide* type strains of each species within the area of the study should be used. These reference organisms may be obtained from culture collections such as the American Type Culture Collection (USA) or the National Collection of Type Cultures (UK).

Once the organisms have been selected, it is essential to confirm purity. Good classifications will not result from use of contaminated cultures. Moreover, it is sound policy to maintain separate reserve and working stocks to protect against accidental loss or contamination. The purity of all cultures should be regularly checked during the course of study. For this, the examination of colonial morphology and Gram-stained smears should suffice.

Finally, it is good practice to duplicate a portion of the strains (approximately 10%) to provide an estimate of the reliability of the data gathering (see section 2.3.2).

Table 2.1 Characters used in numerical taxonomy of bacteria

Category of test	Test
Colonial morphology	Presence of non-diffusible or diffusible, fluorescent, non-fluorescent or luminous pigments. Colony size and shape (presence of spreading colonies).
Micromorphology	Gram staining and acid fast staining reactions. Presence of cocci, bacilli, mycelia, sheaths. Cell size. Attachment structures. Presence of intracellular granules. Motility (polar or peritrichous flagella, or gliding). Evidence of di- or pleomorphism. Presence of spores.
Growth characteristics	Presence of ring or pellicle, or turbid or flocculent growth in broth. Aerobiosis or anaerobiosis. Growth in 0–10% (w/v) sodium chloride. Special requirements for growth, e.g. amino acids, vitamins or metal ions.
Biochemistry	Fermentative or oxidative metabolism of glucose. Presence of catalase, oxidase, and other enzymes. Ability to degrade complex molecules, e.g. starch and tributyrin. Acid production from carbohydrates. Production of indole and H_2S. Methyl red test. Nitrate reduction. Voges–Proskauer reaction.
Inhibitory tests	Presence or absence of growth in the presence of antibiotics, dyes and other inhibitory compounds, e.g. potassium cyanide.
Utilization of compounds as the sole source of carbon for energy and growth	Presence or absence of growth in the presence of carbon containing compounds, e.g. alanine, fructose, maltose and sodium citrate.
Serology	Presence or absence of agglutination reactions to specific antisera.
Chemotaxonomy	Presence or absence of sub-cellular components, e.g. mycolic acids and menaquinones.
Molecular genetics	Presence of specified guanine plus cytosine (G+C) ratios of the DNA.
Phage typing	Presence of specified bacteriophage typing patterns.

2.3   TEST SELECTION

There are no hard and fast rules governing the choice of tests (characters) to be adopted. The possibilities may appear to be endless, although many characters reflect subtle variations on central themes. Essentially, the choice may include representatives of the 'classical' phenotypic tests, such as indole production and the ability to ferment lactose, and the 'newer' methods, encompassing chemotaxonomy, for example presence or absence of specific menaquinones or cell wall sugars (see Chapter 4), or molecular features including DNA homology values (Table 2.1). On balance, it would appear that most numerical taxonomy studies to date have emphasized metabolic rather than chemotaxonomic traits.

The ideal situation would be to use characters representing the expression of single genes or operons that are not subjected to environmental changes. Such characters should be stable, and therefore enable reliable natural classifications to result. Tests, based on single properties, are referred to as unit characters, and by definition comprise taxonomic characters of two or more states, which cannot be subdivided logically, except for changes in the method of coding. Needless to say unit characters vary in the amount of genetic information they represent. The presence of an endospore and the utilization of sucrose are both unit characters, but the former represents some 50 operons whereas the latter just one. With the current awareness of microbial molecular biology, it is impossible to sub-divide sporulation into the constituent characters and, for practical purposes it is considered as a single or unit character. It is over this point that critics of numerical taxonomy and the *a priori* equal weighting of characters are generally most vociferous, but the practical concept of the unit character is well established and is generally regarded as the best solution.

In ideal circumstances, the testing regime should comprise a random assortment of all the possible characters which could be studied. This would preclude the temptation of *a priori* weighting of the characters through choice of specific features. However, in practice, it is satisfactory to use a battery of tests representing as many facets of the biology of the organisms as possible. An ideal list will include colony and micromorphology data, growth characteristics, biochemical tests, effect of potentially inhibitory agents, utilization of compounds as the sole source of carbon for energy and growth, and serological, chemotaxonomic and molecular genetic information (Table 2.1). It is a sensible precaution to use approximately equal numbers of characters from each of the general categories mentioned above, as this will minimize influencing the outcome of the exercise in favour of certain facets, for example classifications based heavily on colonial and micromorphological features. When surveying the tests to be included, it is prudent to select those that are simple and inexpensive in both time and money because a large number of tests will be carried out and complicated time-consuming chemical analyses would not be appropriate.

Redundant tests should be avoided. These include characters that are wholly positive or negative for the OTUs under study, and would, therefore, have no discriminatory value. As an extreme example, a test scoring the presence or absence of DNA in a bacterial cell would contribute nothing meaningful to the classification. Unfortunately uniformly negative or positive tests are often not known until all the tests have been completed! Poorly defined or poorly reproducible tests should also be avoided since ambiguous or wrong answers would result. A fluctuation in recording poorly defined tests is likely to occur, particularly if a large number of OTUs are under study. An example might concern an over-ambitious attempt to define colony colour in terms of slight differences between white, off-white, cream, pale yellow and yellow. The interpretation of such slight colour variations is likely to vary considerably from the start to finish of any testing regime and would introduce an error component into the study.

The optimum number of characters to be studied for each OTU would appear to lie between 100 and 200. Greater numbers have been used by some workers, but the increased information content of the finished study rarely justifies such enthusiasm. Essentially, classifications are not substantially improved by the huge numbers of characters, and the optimum number seems to be 100–150. In contrast, the effective minimum is approximately 50 characters. The obvious danger from smaller batteries of tests is that each of the characters would exert a disproportionately large influence on the analyses, particularly if any errors occurred. However, the problem of test error and, for that matter, test reproducibility will be considered in the next section. As far as possible, each character should be recorded for each OTU because incomplete data may influence the outcome of the computation.

Careful consideration needs to be given to the exact test methods employed. There must be strictly standardized conditions for medium formulation, inoculation procedures, incubation times and temperatures. Any special condition required by one OTU must be applied to all OTUs under study. For example, if one OTU requires 1.0% (w/v) sodium chloride and/or 0.1% (w/v) L-cysteine hydrochloride for growth, then all strains should be exposed to the same concentrations of these compounds. It is desirable that all media should be inoculated with a standard quantity of cells, preferably in the logarithmic phase of growth. Moreover, all strains should be incubated at a common (optimum) temperature for the same period of time. This necessitates the use of sufficient incubation periods to permit the growth of the slowest organism. Where possible, control strains should be incorporated into the study to check on the sensitivity of media and reagents and their reactions.

2.3.1   Rapid methodology and automation

Traditional biochemical tests often require incubation for one or two weeks which hinders classification and, more importantly, identification. The identification schemes which will be developed from the classification should provide reliable and rapid identification of unknown organisms. To achieve this, there has been a move towards rapid miniaturized tests based on commercial kits, such as the API system (described in Chapter 7). These kits, designed primarily for identification purposes, can be used to gather large amounts of taxonomic data in a simple and reliable manner. Characters based on enzyme reactions are even quicker to obtain than the growth-based reactions of the API system. Enzyme substrates, such as nitrophenyl derivatives of sugars which release the yellow compound nitrophenol when hydrolysed, or methylumbelliferyl derivatives which fluoresce when hydrolysed, can be used to gather large amounts of metabolic information about strains very quickly and with high reproducibility. Companies such as Sensititre incorporate numerous enzyme substrates into 96-well microtitre trays which can be inoculated with a suspension of the test organism and incubated in a dedicated incubator / scanner which will read the results after about 4 h. A similar system using a microtitre tray containing 96 substrates and a tetrazolium salt to indicate carbon source metabolism is marketed by Biolog as an identification system, but similarly can be used to gather the characteristics of OTUs for numerical classification. The data collection for numerical taxonomy may therefore be less tiresome than in the past when large numbers of individual tests had to be performed.

2.3.2   Test error and reproducibility

It is sound scientific practice to duplicate experiments in order to detect errors and to assess reproducibility and numerical taxonomy is no exception. Indeed, these aspects are particularly important when examining large numbers of strains for numerous characteristics when it is very easy for mistakes to creep into the data, whether reflecting errors in recording or uncertainty in the correct interpretation of test results. It is noteworthy that freshly isolated environmental strains are particularly capable of losing metabolic activities, such as synthesis or degradation of complex molecules upon repeated sub-culturing in the laboratory. Perhaps this reflects the loss of plasmid DNA or other more subtle changes in the bacterial physiology. Problems abound in the 'correct' interpretation of so-called borderline reactions; is the result weakly positive or negative? As stated previously in studies of large nbers of OTUs, variations in recording occur during the course of study. This problem of test error / reproducibility in numerical taxonomy was voiced initially by Professor Sneath (Sneath and Johnson, 1972) and the outcome was a strong recommendation that an estimate of test error should form an integral part of any numerical taxonomy study. Since duplication of all OTUs in the study would greatly increase the workload, it is common to use randomly picked duplicate cultures, the identity of which is concealed to reduce experimenter bias. These duplicate cultures, amounting to approximately 10% of the total number of OTUs, should be included with the collection of strains and examined throughout the testing regime. On completion, the data from the duplicate cultures are compared and used to estimate test error.

The variances of individual tests between replicate organisms (S_i^2) may be calculated from the equation:

$$S_i^2 = n/2t \qquad \text{(Equation 15; Sneath and Johnson, 1972)}$$

Here, n corresponds to the number of OTUs with discrepancies in the test, and t is the total number of duplicate strains. This may be illustrated by examination of the specimen data set in Table 2.2.

Table 2.2 Specimen data set

	OTU	A	B	C
			Test	
Original	1	+	+	−
Duplicate culture	1	+	−	−
Original	2	+	+	+
Duplicate culture	2	−	+	+
Original	3	+	+	+
Duplicate culture	3	+	−	+

Thus, the individual test variances (S_i^2) among the duplicate strains for tests A, B and C are 0.16, 0.33 and 0, respectively.

The probability of error for an individual test (P) is:

$$P_i = 1/2\,[1-\sqrt{(1-4S_i^2)}] \qquad \text{(Equation 4; Sneath and Johnson, 1972)}$$

For the above example:

$$P_A = 1/2\,[1-\sqrt{(1-4 \times 0.16)}] = 0.2 = 20\%$$

In this way individual tests can be examined for test error.

The individual test variances may be averaged to provide a global estimate of error. The formula is:

$$S^2 = 1/N\,(S_A{}^2 + S_B{}^2 + \ldots S_N{}^2),$$

where N equals the total number of tests, and $S_A{}^2$, $S_B{}^2$ and $S_N{}^2$

correspond to the individual test variances for tests A, B and up to N, respectively. Continuing with the above example:

$$S^2 = 1/3 \ (0.16 + 0.33 + 0) = 0.161$$

The pooled variance (S^2) may then be used to determine the average probability of an erroneous test, i.e. global test error among the data using Equation 4 of Sneath and Johnson (1972). For the above example this is 0.29 or 29%.

In cases where P is greater than 10% (as in the above example), there may be serious problems with the reliability of the numerical taxonomy study, and an erroneous measurement of similarity between the OTUs may result. It is therefore a wise precaution to reject individual tests which show a probability of error of more than 10%. Ironically, some of the classical bacteriological tests, such as nitrate reduction, oxidase, gelatinase and urease production, often appear in this category. However, the evidence points to poorer test reproducibility at the inter- rather than intra-laboratory level. For numerical taxonomy, it should suffice to use large numbers of characters, thereby reducing considerably any detrimental effects caused by a small number of poorly reproducible rogue tests.

2.4   DATA CODING

With completion of the testing regime and the production of neat and comprehensive data books, the next stage is to code the information in a format suitable for computation. A table, referred to as an $n \times t$ table,

Table 2.3   Examples of characters and character states used in the numerical taxonomy of bacteria

Character	Character state
Colony coloration	White / off white / yellow / orange / red / purple
Diffusible fluorescent pigment	Present / absent
Gram-staining reaction	Gram-positive / Gram-negative / Gram-variable
Micromorphology	Rods / cocci / mycelia
Motility	Present / absent
Catalase production	Present / absent
Nitrate reduction	Present / absent
Degradation of starch	Present / absent
Growth on 7% (w/v) sodium chloride	Present / absent
Utilization of individual compounds as the sole source of carbon for energy and growth	Present / absent

is produced, which lists the test results for all the OTUs in the investigation. Unit characters, which exist in either of two states, i.e. present or absent, are coded numerically as 1 and 0 corresponding to + and − respectively (Table 2.3). Some program packages can accept missing or non-comparative data, which are usually coded as either 3 or 9. It must be emphasized that there will be variations in the required data format according to the software used.

Qualitative data may be reduced to two or more multistate characters by non-additive coding. In the example of colony pigmentation, four possible states of three characters have been defined in Table 2.4. Of course, it could be argued that there has been *a priori* weighting of pigmentation by splitting it into three characters and making it three times more important than a simple binary character. Thus a colony that is purple is not, of course, orange or red. This gives these different colony types as much in common as they have to distinguish them. At this level, this does not overly distort the situation but the addition of more colony colours would increase the amount of negative correlation.

Table 2.4 Non-additive coding of qualitative characters

Character	OTU			
	1	2	3	4
Colony red	0	1	0	0
Colony orange	0	0	1	0
Colony purple	0	0	0	1

Data, for which an OTU is assessed quantitatively, may be scored as a series of unit characters in additive coding or directly as numerical values. Cell length commonly comes into this category. In this case, the three characters are again presented as four states (Table 2.5). Again, it could be argued that there has been *a priori* weighting of the characters, but, in fact, the additive coding, in which the longer OTU (OTU4) is considered to be positive for the shorter lengths as well as

Table 2.5 Additive coding of quantitative characters

Character	OTU			
	1	2	3	4
Cell length (µm)				
1–1.9	0	1	1	1
2–3	0	0	1	1
> 3	0	0	0	1

the largest category, retains the dimension of the character and only applies excessive weight if the character has been divided into too many unit characters. In Table 2.5, OTU4 (which is longer than 3 μm) has three differences from OTU1 (which is less than 1 μm in length) and only one difference from OTU3 (which is between 2 and 3 μm long). An example of a quantitative character coded directly as a numerical value would be the colony diameter measured in mm.

The data, in the form of the $n \times t$ table, may now be entered into the computer. Verification of the coded data is a thankless, but important, task which serves to check for mistakes. Assuming that all is well, the data are now ready for computer analyses.

2.5   COMPUTER ANALYSES

Since the initial developmental work in the 1950s, computational procedures have evolved with ever-increasing sophistication. Jargon has become widespread, and it is understandable for the uninitiated to be frightened by the esoteric statistical language. For the present, it suffices to emphasize that the initial role of the computer is to compare the data for each OTU with every other OTU. This enables the level of similarity or dissimilarity (known also as distance) to be calculated between each and every OTU. The most widely used methods in bacterial taxonomy calculate similarity values. Therefore, these techniques will be discussed initially.

2.5.1   Calculation of resemblance

Only a few of the many possible methods for measuring similarity values find routine use in bacteriology. The formulae of some potentially useful methods, referred to as coefficients, are shown in Table 2.6. Quite simply, these coefficients measure the relationships between pairs of OTUs and have been devised with diverse disciplines in mind, including botany, palaeontology and zoology. Nevertheless, they serve essentially the same purpose. From Table 2.6, it will be observed that these formulae are based on a group of symbols; a, b, c and d. These correspond to the number of positive (a) and negative (d) matches, and the number of dissimilar results (b, c) between a pair of OTUs. This relationship may be represented diagrammatically as:

		Results for OTU1	
		+	−
Results for OTU2	+	a	b
	−	c	d

Table 2.6 Some coefficients with useful discriminating value for microbiology (after Austin and Colwell, 1977)

Coefficient	Abbreviation	Formula
Simple matching	S_{SM}	$\dfrac{a + d}{a + b + c + d}$
Jaccard	S_J	$\dfrac{a}{a + b + c}$
Dice	S_D	$\dfrac{2a}{2a + b + c}$
Hamann	S_H	$\dfrac{(a + d) - (b + c)}{a + b + c + d}$
Kulczynski	S_{K2}	$\frac{1}{2}[\{a / (a + b)\} + \{a / (a + c)\}]$
Ochiai	S_O	$\dfrac{a}{\sqrt{[(a + b)\,(a + c)]}}$
Pattern difference	D_P	$\dfrac{2\,\sqrt{(bc)}}{a + b + c + d}$
Rogers and Tanimoto	S_{RT}	$\dfrac{a + d}{a + d + 2(b + c)}$
Total difference	D_T	$\dfrac{b + c}{a + b + c + d}$
Unnamed	S_{UN1}	$\dfrac{2(a + d)}{a + b + c + d}$
	S_{UN4}	$\frac{1}{4}[\{a / (a + b)\} + \{a / (a + c)\}\{d / (b + d)\} + (d / a)]$
Angular transformation of $S_{SM} \times 2/\pi$	$\mathrm{Sin}^{-1}(S_{SM})$	$0.637 \times \arcsin\left(\sqrt{\dfrac{a + d}{a + b + c + d}}\right)$

a and *d* correspond to the number of positive and negative matches, respectively.
b and *c* represent the number of non-matching characters between pairs of OTUs.

To illustrate this calculation, let us examine a sample data set:

Test	OTU 1	OTU 2	OTU 3
1	1	1	0
2	0	0	1
3	1	0	1
4	0	0	0
5	1	1	1
6	0	0	0
7	1	1	1
8	1	1	1
9	1	0	0
10	0	1	1

For this 10 test data set, the number of positive matches, (the *a* component) between the pair of OTUs 1 and 2 is 4. Similarly, the number of negative matches (the *d* component) is 3. The number of tests for which OTU1 is negative but OTU2 is positive (the *b* component) is 1, and the converse of *b* (the *c* component) is 2. Similarly, for the comparison OTU1 and OTU3, $a = 4$, $d = 2$, $b = 2$ and $c = 2$, and for the comparison of OTUs 2 and 3, the values are 4, 3, 2 and 1, respectively. The summation of $a + b + c + d$, is the total number of tests, which in this example is 10.

One of the most widely used coefficients is the simple matching coefficient (S_{SM}), which measures positive matches (a) and negative matches (d) as a proportion of the total number of characters $(a + b + c + d)$. Thus, $S_{SM} = (a + d)/(a + b + c + d)$. For our example, S_{SM} for OTUs 1 and 2 is $7/10 = 0.7$ or 70%; for OTUs 1 and 3 = 60% and for OTUs 2 and 3 = 70%. The scale of values with this coefficient will range from 0, for total dissimilarity, to 1 for total similarity between pairs of OTUs.

A second coefficient, which finds widespread use is the Jaccard coefficient (S_J). This coefficient discounts negative matches. Thus, $S_J = a/(a + b + c)$ which for our example works out as $4/7 = 0.57 = 57\%$ for OTUs 1 and 2 and for OTUs 1 and 3 = $3/8 = 0.375 = 37.5\%$ and for OTUs 2 and 3 = $4/7$ or 57%. The range of values also spreads from zero to unity and it is straightforward to convert the values into percentages. The S_J and S_{SM} coefficients have been used in most publications concerning numerical taxonomy studies of bacteria. It will be readily apparent that differences in similarity values between OTUs will result according to the nature of the coefficient used and will invariably be lower using S_J compared with S_{SM}. This begs the question about which is the 'correct' method to be used, since the choice of coefficient will affect the outcome of classification. This dilemma will be addressed later.

Another coefficient, which has been used occasionally in bacteriology, is Gower's coefficient (S_G). This is a weighted average of all similarity values between pairs of OTUs and is suitable for use with binary, quantitative and qualitative data. For binary data S_G is equivalent to the S_J coefficient, whereas for qualitative characters of two or more states S_G is identical to the S_{SM} coefficient. However for quantitative data:

$$S_G = \left[1 - \frac{\text{(the value of OTU1 for a given character minus the value of OTU2 for the same character)}}{\text{the range of values for the character}} \right]$$

A fairly complex example may be illustrated by use of a sample data set (Table 2.7).

Table 2.7 Sample data set

OTU	Motility	Colony diameter (mm)	Guanine plus cytosine (G + C) ratios of the DNA (mol%)	Colony colour
1	+	2.5	48	Orange
2	+	2.3	62	Yellow
3	−	1.8	43	Cream
4	−	1.5	42	Cream

The range of values for colony diameter, i.e. 1.5–2.5 mm, is 1.0. For the G+C ratio, the range, i.e. 42–62 mol %, is 20. Dealing with the calculation for S_G of OTU1 and OTU2, it may be determined that:

$$S_G = \frac{\overset{\overset{\text{i}}{\downarrow}}{1} + \overset{\overset{\text{ii}}{\downarrow}}{(1 - 0.2/1.0)} + \overset{\overset{\text{iii}}{\downarrow}}{(1 - 14/20)} + \overset{\overset{\text{iv}}{\downarrow}}{0}}{1 + 1 + 1 + 1} = 0.525$$

Here i, ii, iii and iv refer to the presence of motility, colony diameter, G+C value, and colony colour, respectively. For iv, the zero score on the top line refers to the non-match of colour between the OTUs. The 1 + 1 + 1 + 1 score for the bottom line refers to the presence of four characters. Use of S_G for OTUs 3 and 4 shows that:

$$S_G = \frac{\overset{\overset{\text{i}}{\downarrow}}{0} + \overset{\overset{\text{ii}}{\downarrow}}{(1 - 0.3/1.0)} + \overset{\overset{\text{iii}}{\downarrow}}{(1 - 1/20)} + \overset{\overset{\text{iv}}{\downarrow}}{1}}{0 + 1 + 1 + 1} = 0.883$$

It may now be obvious that the 'i' component has been scored as zero because of the absence of motility. However, colour 'iv' is scored as unity because of the match, i.e. both OTUs produce cream colonies.

Obviously, there is a requirement for many arithmetic calculations to compare data for all the OTUs in the study. In fact, the total number of calculations between OTUs is $N = \frac{1}{2}n\,(n-1)$, where n is the total number of OTUs. For groups of 50, 100 and 200 OTUs, it would be necessary to carry out 1225, 4950 and 19 000 calculations, respectively. Thus, there is an exponential increase in the number of calculations commensurate with a linear increase in the number of OTUs.

2.5.2   Distance coefficients

Both novice and experienced numerical taxonomists may express reservation about the terminology referring to dissimilarity measurements. These measurements, which are also referred to as metrics and distance *(D)*, are the converse of, and are complementary to, methods of assessing similarity. In this section, discussion will centre around Euclidean distance, which has been used in numerical taxonomy studies of bacteria. An excellent description of the dissimilarity measurement is to be found in Dunn and Everitt (1982). It is important to emphasize that this measurement satisfies the conditions of Pythagoras's theorem, which concludes that the squared value of the hypotenuse of a right-angled triangle equals the sum of the squared values of the other two sides. In brief, Euclidean distance measures the distance between OTUs in terms of their co-ordinates on right-angled (also referred to as Cartesian) axes. The values extend from zero to an infinitely large amount depending upon the number and size of the differences between OTUs. Therefore, these values may not be trans-posed to percentages. As Euclidean distance is determined from Pythagoras's theorem, D^2 is calculated, which is itself used as a measure-ment of dissimilarity. The advantage of using D^2 is that it avoids the necessity of calculating square roots. In its simplest form for binary data, D corresponds to $\sqrt{(1-S_{\mathrm{SM}})}$ and gives the diagonal distance between two OTUs. However, in this case, use of D^2 is beneficial insofar as the dissimilarity between OTUs corresponds to the number of binary characters by which they differ, i.e. the b and c component of the previously mentioned notation. This assumes that missing data are not present in the data matrix. Therefore in these circumstances $D^2 = (b + c)/(a + b + c + d)$, whereas $D = \sqrt{(b + c)/(a + b + c + d)}$. For quantitative multistate characters, the Euclidean distance between a pair of OTUs is defined as:

$D = \sqrt{}$(the value of the first OTU for a given character minus the value of the second OTU for the same character)^2

This calculation continues by adding similar information for the second character, continuing until all of the characters have been so considered. Euclidean distance may be expressed as (see, for example, Clifford and Stephenson, 1975):

$$D = \left[\sum_{1}^{n} \left(\chi_{OTUA} - \chi_{OTUB} \right)^2 \right]^{1/2}$$

where Σ is the summation sign for all the characters from the first, i.e. 1 to the last, i.e. nth; and χ_{OTUA} and χ_{OTUB} correspond to the value of OTUA for χ and the value of OTUB for the same character, respectively. The determination of Euclidean distance for quantitative data may be illustrated, using the following example:

OTU	Length of cells (µm)	G+C ratios of DNA (mol %)	Great quantity of sodium chloride permitting growth (%)
1	2.0	42	1.5
2	2.5	45	2.0
3	3.0	59	6.0

For OTU1 compared with OTU2:

$$D^2 = (2.0 - 2.5)^2 + (42 - 45)^2 + (1.5 - 2.0)^2 = 9.5$$

$$D = 3.08$$

For OTU1 measured against OTU3:

$$D^2 = (2.0 - 3.0)^2 + (42 - 59)^2 + (1.5 - 6.0)^2 = 310.25$$

$$D = 17.61$$

Similarly for OTU2 assessed against OTU3:

$$D^2 = (2.5 - 3.0)^2 + (42 - 59)^2 + (2.0 - 6.0)^2 = 212.25$$

$$D = 14.56$$

From this example, it is apparent that OTU1 is closer to OTU2 than to OTU3.

2.5.3 Vigour and pattern statistics

A further significant development by Sneath in 1968 concerned the effect on bacterial taxonomy of vigour and pattern elements. It was reasoned that the total difference between OTUs reflected two components, namely pattern differences and vigour differences, which may be equated with shape and size difference, respectively, in animals or plants. The object of Sneath's work was to reduce the apparent dissimilarity between OTUs that could be attributed to differences in growth rate, i.e. the vigour difference. Slow-growing strains may be recorded as negative in a test, such as acid production from sugars, not because they lack the necessary catabolic pathway but because they do not grow fast enough for a positive reaction to be recorded within the specified time period. This coefficient is, therefore, useful when strains of widely different metabolic activity are being studied. The vigour of an organism was defined as the proportion of its characters that showed a positive response in the tests examined, and the difference in vigour (D_v) is simply the difference in this value between two OTUs. According to Sneath, there is a relationship between vigour and pattern, as follows (using our previous notation):

$$D_v = c - b/a + b + c + d$$

Sneath partitioned total difference (D_t) as the sum of differences in vigour and pattern as shown below:

$$D_T^2 = D_v^2 + D_p^2$$

From this equation and that for D_v, the pattern difference (D_p) can be derived as:

$$D_p = 2\sqrt{(bc)}/a + b + c + d$$

D_p ranges in value from zero for identical pairs of OTUs to unity for completely dissimilar pairs of OTUs and has been useful for studies of bacteria that grow weakly, particularly when comparisons are to be made with more vigorous organisms (e.g. Goodfellow *et al.*, 1976).

The outcome from all the similarity or distance calculations, whatever the coefficients used is an unsorted (similarity or dissimilarity) matrix that records the similarity/distance between each of the OTUs. In practice, this is a triangular array, and approximates to mileage charts showing distances between towns (see Fig. 2.3).

2.6   DETERMINATION OF TAXONOMIC STRUCTURE

Having calculated the relationships between all OTUs, the next stage is to order the organisms into groups of high overall similarity (or phena) which will form the species and genera of our classification. Essentially, this can be done in one of two ways. Hierarchical procedures produce a ranked classification in which the strains group into species, species into genera, genera into families etc. This is the common form of classification with which most of us are familiar. Ordination procedures, on the other hand, group strains into taxa but do not indicate how these taxa might be organized into higher ranks. Thus, species might be indicated but not genera. However, before we embark

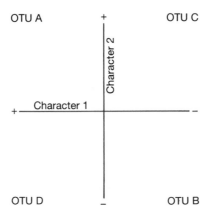

Fig. 2.2 Representation of a two-dimensional character or attribute (A-space). Organisms can be assigned to one of four classes defined by two characters and here represented by four OTUs (A, B, C and D). A third character requires a third dimension at right angles to the previous two to form a cube and so on for further characters (see text for details).

on our consideration of taxonomic structure, it will be useful to provide some comment on the concept of taxonomic space.

2.6.1 Taxonomic space

The relationships of OTUs are configured in a taxonomic space. In essence, this means that OTUs are placed in a space of which the coordinates are the characters of the OTUs. For two characters this can be readily conceived as shown in Fig. 2.2 in which four OTUs (A, B, C and D) have been placed in a two-dimensional plane according to their characteristics (OTUA is positive for both characters, OTUB negative for both and OTUs C and D have opposite positive and negative characteristics). This is an accurate representation of the relationships of the four OTUs. If a third character was added, a third dimension would be necessary which should be at right angles to the existing axes. This would form a cube. As more characters are added, so more axes are added, and the governing feature is that each should be at right angles to the existing axes. Of course this cannot be visualized but it can be represented mathematically, and is described as a multidimensional 'attribute' or A-space. A second taxonomic space termed an 'individual' or I-space can be similarly constructed, in which the OTUs are the axes and the characters are plotted; but this is not of immediate relevance here.

In creating our taxonomic structure, OTUs will be placed in the multidimensional A-space according to the values of the characters. We will then define groups or clusters of OTUs in the taxonomic space

Data (similarity) matrix

```
OTU
  1│100
  2│ 60 100
  3│ 70  90 100
  4│ 80  70  60 100
  5│ 40  30  40  50 100
   └─────────────────────
     1   2   3   4   5
```

Results of clustering

% Similarity	Single linkage	Average linkage	Complete linkage
100	1 2 3 4 5	1 2 3 4 5	1 2 3 4 5
90	2,3 1 4 5	2,3 1 4 5	2,3 1 4 5
80	2,3 1,4 5	2,3 1,4 5	2,3 1,4 5
70	2,3,1,4 5	2,3 1,4 5	2,3 1,4 5
60	2,3,1,4 5	2,3,1,4 5	2,3,1,4 5
50	2,3,1,4,5	2,3,1,4 5	2,3,1,4 5
40	2,3,1,4,5	2,3,1,4,5	2,3,1,4 5
30	2,3,1,4,5	2,3,1,4,5	2,3,1,4,5

Dendrograms

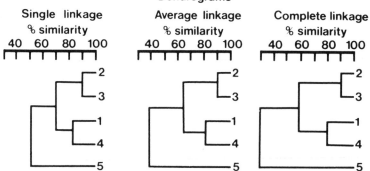

Single linkage % similarity	Average linkage % similarity	Complete linkage % similarity

Cophenetic correlation matrices

```
1│100      r=0·90        1│100      r=0·90        1│100      r=0·90
2│ 70 100                2│ 65 100                2│ 60 100
3│ 70  90 100            3│ 65  90 100            3│ 60  90 100
4│ 80  70  70 100        4│ 80  65  65 100        4│ 80  60  60 100
5│ 50  50  50  50 100    5│ 40  40  40  40 100    5│ 30  30  30  30 100
  ─────────────────        ─────────────────        ─────────────────
   1   2   3   4   5        1   2   3   4   5        1   2   3   4   5
```

Fig. 2.3 Process of single, complete and average linkage cluster analysis. A comma between two strain numbers denotes that the two strains are included in a cluster (see text for details).

which will become the species and genera. To view the output of this procedure, it must be reduced to two dimensions in the form of a graph or diagram, or three dimensions as a space-filling model. By necessity, this reduction from about 100 dimensions to just two will introduce distortion into the classification, and it is our aim to reduce this distortion to the minimum by using appropriate clustering algorithms. The distortion will be less evident at the higher similarity levels, at which OTUs join into species, than at the lower levels where clusters fuse into higher-ranked taxa such as genera. It is therefore unwise to make taxonomic judgements at lower similarity levels in a hierarchical classification, particularly where generic or family designations might be introduced.

2.6.2   Hierarchical clustering

The ordering of OTUs into groups of high overall phenetic similarity is usually achieved by means of one of several commonly used hierarchical clustering methods. Popular procedures include single- and average-linkage clustering. In addition, there is limited interest among bacterial taxonomists for complete-linkage, centroid sorting and Ward's method. The relative values of these methods will be considered later.

In the case of the linkage methods, the algorithm starts by searching the similarity matrix for the highest value between any pair of OTUs. The similarity between these OTUs is then listed. This pair(s) with the highest similarity value is treated as a single OTU. The computer then computes the next highest pair, which may be between two other OTUs or the previous pair and another OTU. The cycle proceeds and at each stage OTUs are added to clusters or clusters joined until all the OTUs are included in a single cluster ($T-1$ cycles for T OTUs).

The clustering techniques differ in the definition of the point at which OTUs join groups and groups coalesce. In single linkage (nearest neighbour) clustering, OTUs join groups at the highest similarity between the OTU and any one member of the group with no account being taken of other members of the group. Thus in the example shown in Fig. 2.3, all five OTUs are distinct at the 100% similarity level. At 90%, OTUs 2 and 3 form a cluster, but OTUs 1, 4 and 5 remain outside the cluster. The next highest similarity at which two OTUs join is 80%, representing OTUs 1 and 4. These two strains therefore form the nucleus of a second cluster. The two clusters fuse at 70% similarity because at least one OTU from each cluster forms a pair at this similarity level. Although the pairs OTU1,2 and OTU3,4 are each only 60% related, nevertheless they are forced into the combined cluster by single linkage analysis by the pairs OTU1,3 and 2,4 which are both 70% related. Remember, it is the highest similarity which dictates the combination of groups. Similarly, OTU5 joins cluster 2,3, 1,4 at 50% similarity because it shares this with OTU4, despite it possessing a lower similarity with the other members of the cluster.

Complete linkage clustering (furthest neighbour) is the antithesis of single linkage clustering and incorporates OTUs into clusters or combines clusters at the lowest similarity level between the OTU and any member of the cluster. Thus, in the example in Fig. 2.3 the two initial clusters formed by the most highly related strains are again formed. However, clusters 2,3 and 1,4 coalesce at 60%, the values for the pairs 3,4 and 1,2, respectively, despite the higher similarity of pairs 1,3 and 2,4. Moreover, OTU5 joins the cluster at 30% similarity, the value for the pair 5 and 2 which is the lowest similarity displayed by OTU5 to any member of the cluster. Thus complete linkage clustering exaggerates the differences between clusters while single linkage clustering smooths out these differences.

Average-linkage, or to give it its full name, unweighted pair group method with arithmetic averages (UPGMA) clustering is a compromise of the single and complete linkage algorithms. In this procedure, an OTU joins a group at the average similarity between the OTU and all members of the group. Again, referring to the example in Fig. 2.3, the initial stages of the clustering process are the same as before and the groups represented by the most highly related OTUs are again formed. The two groups 2,3 and 1,4 fuse at the mean similarity of the group individual relatedness values; 2,1 = 60%; 2,4 = 70%; 3,1 = 70%; 3,4 = 60% which is 60 + 70 + 70 + 60 = 260/4 = 65%. OTU5 joins the cluster 2,3,1,4 at the average similarity of it with all members of the cluster which is 40 + 30 + 40 + 50 = 160/4 = 40%. Although it takes longer in computer time, it can be shown theoretically that average-linkage cluster analysis gives the most accurate representation of the taxonomic structure and introduces the least distortion into the low-dimensional taxonomic space.

Centroid sorting/clustering joins pairs of OTUs (or an OTU into a group of OTUs) according to the coordinates of their centroids (average value) when considered in multidimensional terms. Groups are joined at the minimum value/distance between the centroids. One major disadvantage of this method is that the identity of small groups of OTUs may be lost upon merging with a large group. In this case, the new centroid position may now fall entirely within the area delineated by the larger group. However, it may be necessary to maintain the relative identity of the original smaller group. This problem was resolved by Gower, who described the technique of median sorting. With this, groups are considered to occupy a unit size, for which an unweighted median position is obtained after fusion.

The fifth clustering method, referred to previously, is Ward's method. With this technique, clustering is based on the minimum sum of squares within clusters.

2.6.3 Ordination methods

Apart from cluster analyses, which account for most numerical taxonomy studies of bacteria, OTUs may be ordered on the basis of two- or three-dimensional arrays. These are ordination diagrams (also known as taxonomic maps), which are non-hierarchical and constitute the so-called multivariate analyses. With such methods it is possible to view the relationship between OTUs directly in terms of taxonomic (phenetic) space. It is important to emphasize that ordination methods (the term was coined by the botanist, Goodall, in 1953) do not divide OTUs into convenient groups, as may be achieved by cluster analyses, although groups may be recognized by careful examination of the output. Instead, ordination methods concern establishing appropriate representations between OTUs in terms of distance and space. Thus, these methods are useful for the examination of entities in which the relationship between them represents a continuum, such as might occur with some groups of bacteria, rather than the more artificial discrete-like 'boxes' highlighted by traditional taxonomy (see Alderson, 1985).

Ordination methods have the potential to be more informative, in terms of relationships between OTUs, than dendrograms. The former usually provide a realistic measure of distance between principal groups, although such distances may be extremely artificial between close neighbours. It is apt to consider the three-dimensional arrangement of planets in space as a parallel to this aspect of bacterial taxonomy. Of course the arrangement will largely depend on the position from which the observations are made. This should be borne in mind when considering ordination methods.

A method, in regular use, is principal component (PCP) analysis, which was first proposed by Pearson in 1901. The guiding concept of PCP is to reduce the multidimensional aspect of the distribution of OTUs in the A-space to just two dimensions so that the positions of the OTUs can be visualized. In moving from some 100 dimensions to two, PCP retains the distinctness of the groups by calculating those two dimensions or components which retain the maximum difference between the OTUs. In simple terms, the relative positions of the OTUs, as measured in terms of distances from each other, are plotted on right-angled (Cartesian) coordinates (axes). A straight line may then be drawn through the position of the OTUs, passing through the centroid (the point representing the centre of the cluster). The location of as many OTUs as possible should be on the line, and the rest are in equal numbers above and below the line (Fig. 2.4).

Technically, it may be described that the position of the line is such that the summation of perpendicular measurements from the pairs of coordinates for each OTU to the line is the minimum value (Clifford and Stephenson, 1975). Such a line between the x and y axes, is shown

Numerical taxonomy

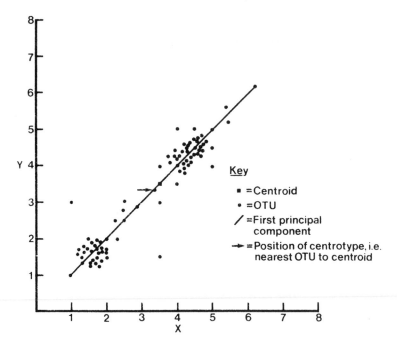

Fig. 2.4 Example of principal components analysis. The line represents the first principal component and is the axis which retains the greatest scatter of spread of the OTUs. Here it groups most of the OTUs into two taxa.

in Fig. 2.4. This line, along which the maximum number of OTUs is spread, is referred to as the first principal component. The next dimension which expresses the next greatest scatter of OTUs is then sought. Having determined the two principal components in this way, these are used as the axes of a graph on which the positions of the OTUs are plotted. The output is often described as a scatter diagram. From these diagrams (such as that shown in Fig. 2.8), it will be evident that, although OTUs can be ascribed to groups (in this case species) there is no indication of higher taxa such as genera. For more detail of how to extract principal components from dissimilarity matrices, the interested reader is referred to the excellent text of Dunn and Everitt (1982).

A similar technique is principal coordinate analysis (PCO), which is essentially the same as PCP but the components are calculated directly from the dissimilarity matrix. Whereas PCP is only relevant when the dissimilarities are based on a Euclidean metric, PCO is applicable whether the distances are Euclidean or not. Indeed, if the distances are Euclidean the results of PCO analysis are equivalent to those of a PCP analysis.

2.7   PRESENTATION AND INTERPRETATION OF RESULTS

2.7.1   Representation of the classification

The results of the classification can be represented in several ways. The sorted similarity matrix has much to recommend it, but is not commonly used. The sorted similarity matrix is constructed by listing the OTUs according to the output from the cluster analysis, i.e. the sorting is provided by the clustering procedure. For example, from the cluster analysis presented in Fig. 2.3, the order for the sorted similarity matrix would be OTUs 2,3,1,4,5 and not 1,2,3,4,5 as in the original similarity matrix. Having sorted the order according to the cluster analysis, the next stage is straightforward (Fig. 2.5). The decimal output from the computer is converted into percentages (Fig 2.5(b)). The numerical values are then converted to shaded diagrams (Fig. 2.5(c) and 2.5(d)), in which different intensities of shading are used to represent bands of similarity values. It is usual for the most and least intense shading to correspond with the highest and lowest similarity values, respectively. In Fig. 2.5(c), values up to 80% and above 90% have been banded for convenience in groups of ten percentage points. Between 81 and 89%, two divisions for the shading have been used. Interpretation of shaded diagrams is much easier than the numerical values expressed in Fig. 2.5(a) and 2.5(b). The clusters are apparent in the sorted similarity matrix as triangles of dark shading on the hypotenuse of the matrix. Five such triangles are apparent in Fig 2.5(d). However, the advantage of the sorted similarity matrix is that it shows areas of overlap between taxa, such as the considerable affinity between cluster 2 and 3 and the high similarity of OTUs 2 and 7 in Fig. 2.5(d). This would probably be hidden by other representations of the results.

Cluster analysis also generates the more popular dendrograms which may be printed directly via graph plotters or constructed manually from salient details. The dendrogram for the OTUs shown in the sorted similarity matrix is shown in Fig. 2.6. The dendrogram shows clusters of highly related OTUs as lines joined together at high similarity levels. For example, OTUs 23 and 24 in Fig. 2.6 form the nucleus of cluster 1 by joining at almost 100%. OTU1 (89%) and OTU18 (85%) are also included in cluster 1. The dendrogram shows the relationships of the strains more clearly than the sorted similarity matrix, but obscures the close similarity between OTUs 2 and 7 which was evident in the matrix. The dendrogram is often simplified by grouping together OTUs clustering together at pre-determined levels of similarity. These groups are presented as shaded triangles, in which the length of the side beneath the 100% similarity level is proportional to the number of OTUs (Fig. 2.6(b)).

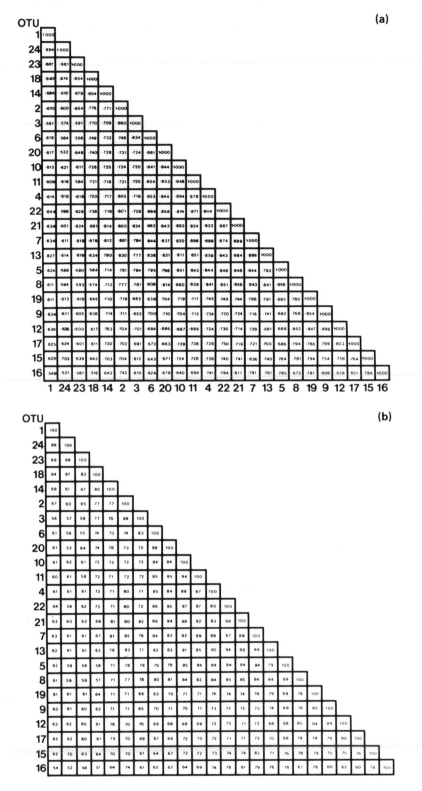

Fig. 2.5 Stages in the production of a sorted similarity matrix. The initial information, to three decimal places of accuracy (a), is converted to percentage values (b), and then to shaded diagrams (c, d). The relatedness between OTUs is readily apparent

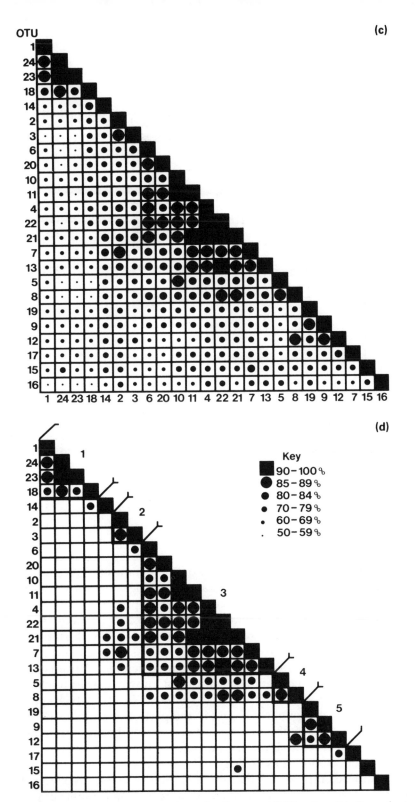

(c)

(d)

Key

■	90–100 %
●	85–89 %
●	80–84 %
•	70–79 %
·	60–69 %
.	50–59 %

by elimination of the lower values from the diagram (d). Numbers 1 to 5 are cluster numbers and the clusters are evident as triangles of dark shading on the hypotenuse of the matrix.

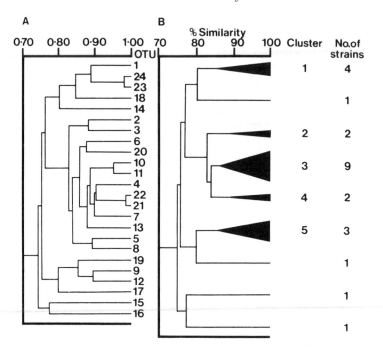

Fig. 2.6 A dendrogram of the classification shown in Fig. 2.5 produced by computer-based graphics (a) may, in turn, be simplified by reducing clusters to triangular areas of shading (b).

The diagrams, illustrated by Figs. 2.5 and 2.6, permit groups/clusters of related OTUs to be recognized. These groups, defined on the basis of overall phenetic similarity, are termed phenetic groups or phena. In Figs. 2.5(d) and 2.6(b), these phena have been labelled 1 to 5. It is generally accepted that species can be defined at the 80–85% similarity (S) level if using S_{SM}–UPGMA generated dendrograms. All strains that coalesce above this level may be considered as presumptive members of a species; although confirmation by chemotaxonomic methods such as DNA reassociation is desirable. Therefore, in Fig. 2.6, clusters 1 to 5 could be considered as five different species. To assist in the recognition of these species it is normal to have included reference strains in the study. The allocation of reference strains of known species to clusters usually indicates the most appropriate similarity level at which species-ranked clusters are being recovered in the particular study.

Similarly, a generic boundary can sometimes be drawn across a dendrogram at 60–65% similarity, but again this needs to be substantiated by chemotaxonomic evidence.

It should be emphasized that these similarity levels are somewhat arbitrary, having been delineated as a result of (numerous) past

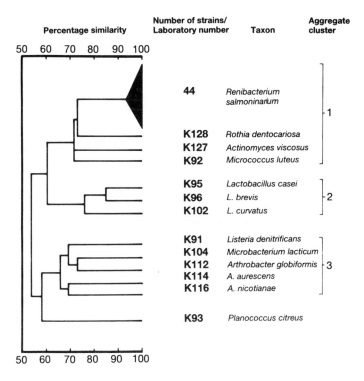

Fig. 2.7 A dendrogram showing the relationship between *Renibacterium salmoninarum* and other Gram-positive organisms based on the S_{SM} coefficient and unweighted average linkage and clustering (from Goodfellow *et al.*, 1985).

experiences of comparing the output of numerical analyses with those of conventional classifications. Of course the wisdom of such action is debatable, and, to some extent, undermines the critical basis of numerical taxonomy. However, the existing evidence, particularly for the Enterobacteriaceae, would support these similarity levels for the definition of species and genera. For other groups, such as the *Flavobacterium – Cytophaga – Flexibacter* group of bacteria, it is difficult to make such generalizations. Yet, the dendrogram can provide a clear and accurate picture of the taxonomic structure of a group of organisms, as shown in Fig. 2.7 which displays the affinities of a Gram-positive fish pathogen, the causative agent of bacterial kidney disease in salmonid fish, with some suspected close relatives. It was originally thought that this bacterium might be closely associated with *Listeria*, but this small numerical taxonomic study showed that *Renibacterium salmoninarum* strains formed a discrete cluster of species status, which was distinct from the reference strains and was appropriately placed in its own

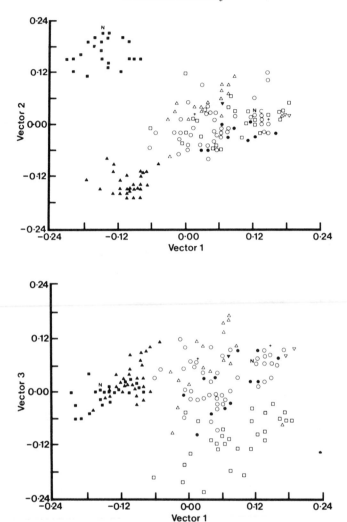

Fig. 2.8 Plot of the first three vectors, accounting for 38% of the total variation, from a principal coordinate analysis of some *Bacillus* strains. *Bacillus anthracis,* ■; *B. cereus,* ○; *B. cereus* strains from diarrhoeal-type food poisoning outbreaks, ●; *B. cereus* strains from emetic-type food poisoning outbreaks, and serotypes 1, 3, 5 and 8, ▲; *B. cereus* var. *fluorescens,* ▽; *B. cereus* var. *mycoides,* △; *B. praussnitzi,* +; *B. thuringiensis,* □; *B. filicolonicus,* ▼; neotype strain, N (from Logan and Berkeley, 1981).

genus. This conclusion was later substantiated by chemotaxonomic studies.

An example of a classification based on PCO analysis is shown in Fig. 2.8. The *Bacillus anthracis* group of organisms comprises several closely related species notably; *B. anthracis, B. cereus* and *B. thuringiensis*.

Moreover, *B. cereus* may be divided into several pathogenic types which cause diarrhoea or vomiting as well as non-pathogens. It is obviously important to be able to recognize these bacteria, particularly *B. anthracis*, the causative agent of anthrax in man and animals. A plot of the first two principal coordinates (Fig. 2.8) distinguished *B. anthracis* and the emetic forms of *B. cereus* from the other organisms in the study, although it failed to separate the diarrhoeal forms of *B. cereus* from *B. thuringiensis*. This is strong evidence for the allocation of the emetic types of *B. cereus* to a separate species and this has recently been substantiated by DNA hybridization studies. Thus PCO analysis was able to distinguish these taxa and was later developed into an identification system for these important pathogens.

2.7.2 Choice of statistical analysis

The outcome of any numerical taxonomy study will reflect the nature of the methods used. Perhaps a legitimate criticism is the lack of standardization in methods, starting with the initial choice of OTUs and the numbers and types of characters to be used. Thereafter, the wide range of computer-based analyses may be confusing. The effect of similarity coefficients on numerical taxonomy studies of bacteria has already been addressed (see Austin and Colwell, 1977). Such differences between the use of the S_{SM}, S_J, S_{K2} and S_{UN4} coefficients (see Table 2.6 for the formulae) for Enterobacteriaceae representatives have been highlighted in Fig. 2.9. It would be pertinent to enquire which (if any) is correct. Unfortunately, there is not a simple answer to this problem. However, most numerical taxonomy studies of bacteria involve use of the S_J and/or S_{SM} coefficients, which appear to generate acceptable results. We will not debate the often heated arguments as to which of these coefficients is best. Suffice to say, the S_{SM} coefficient usually enables the delineation of more clearly defined clusters of OTUs.

Conversely with the S_J coefficient, clusters are more diffuse, with OTUs joining together at much lower S-levels. Yet, because this coefficient ignores negative matches, it is particularly useful in comparing unreactive organisms, such as *Flexibacter* spp., which might otherwise cluster together on the basis of characteristics which the OTUs did not possess, i.e. negative correlation. Consequently, the S_J coefficient is useful for confirming the validity of clusters defined initially by use of the S_{SM} coefficient. Routinely, use of these two coefficients together, perhaps with vigour and pattern statistics should suffice. The latter sometimes highlights seemingly misplaced OTUs, by permitting their recovery in natural groupings.

Again the choice of clustering algorithm will influence the classification. The effect of different cluster techniques on a data matrix for

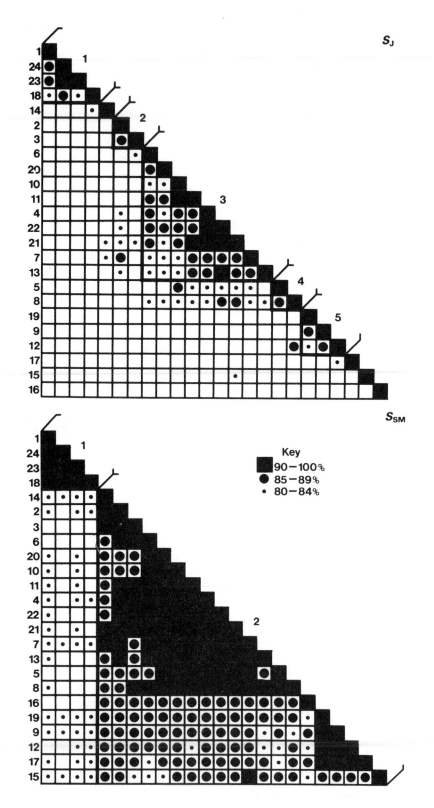

Fig. 2.9 The effect of different similarity coefficients on data for Enterobacteriaceae representatives. Clustering is by the unweighted average linkage algorithm.

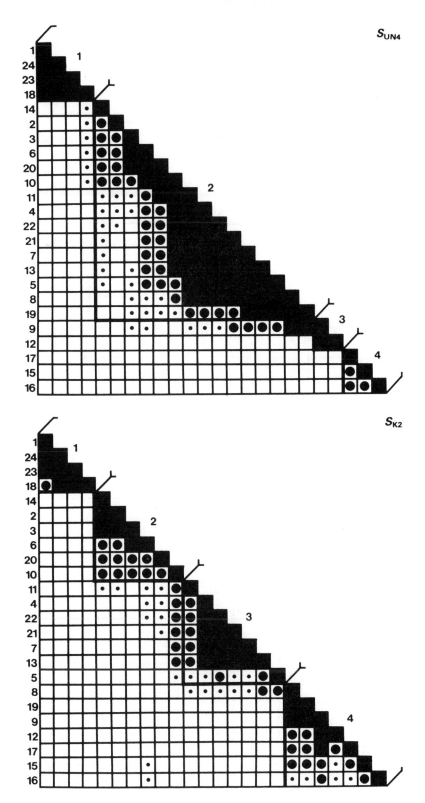

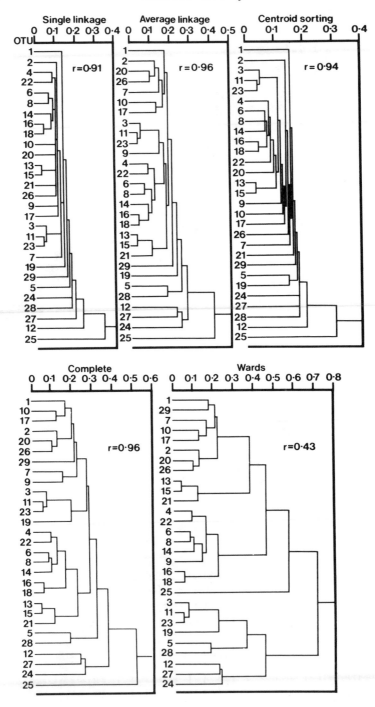

Fig. 2.10 The effect of clustering methods on a data matrix for *Bacillus* isolates, examined by squared Euclidean distance (r = cophenetic correlation; see text for details).

Bacillus strains, which was derived using squared Euclidean distance, is shown in Fig. 2.10. There are pronounced differences in the classifications. Complete linkage and Ward's method produced the most clearly defined clusters, but this was accompanied by increased distortion of the taxonomic structure. OTUs were being assigned to clusters when they would be more correctly recovered as single member clusters. At the other extreme, single linkage and centroid sorting resulted in considerable chaining and few distinct clusters. It can be shown theoretically that UPGMA clustering provides the least distortion in a dendrogram and is usually the clustering algorithm of choice.

2.7.3   Evaluation of results

The accuracy with which a dendrogram represents a similarity matrix can be assessed using the determination of cophenetic correlation. In this procedure, a similarity matrix is generated from the dendrogram. In this matrix, the similarity figures are extracted from the dendrogram as the highest similarity levels linking two OTUs (see Fig. 2.3). This second matrix is then compared with the first using the Pearson product–moment correlation coefficient (r). A figure of $r = 1.0$ would be ideal, and would indicate that all the similarity levels in the dendrogram linking strains are identical to those in the original similarity matrix. However, since all two-dimensional representations of taxonomic structure (which is multidimensional) must introduce distortion, this optimum cannot be obtained. Average-linkage (UPGMA) cluster analysis generally gives the highest cophenetic correlation (Fig. 2.10; the example given in Fig. 2.3 is not representative because of the small sample of strains).

A second approach to the evaluation of a dendrogram is the determination of intra- and inter-group similarity values. These are the average similarities between OTUs within and between groups, respectively. High intra-group similarity levels indicate tight, homogeneous clusters and low inter-group similarity levels show that the groups are distinct. These are the optimum requirements of a numerical classification and would show that the OTUs have been recovered in discrete, clearly distinguished phena. Other computer software is available (see Sackin, 1987), which calculates 'overlap' between clusters. This is an estimation of the disjunction index for each pair of clusters and the corresponding 'nominal overlap'. Generally, one would expect the overlap between pairs of clusters to be less than 5%. Failure to achieve this would suggest that the clusters appear distinct on the dendrogram but actually are not clearly separated. This will have implications for identification, since it will be difficult to determine characters which define clusters if the overlap between clusters is too great.

Numerical taxonomy

2.7.4 Presentation of phenon descriptions

Clusters of related OTUs (phena) may be further characterized by tabulating the percentage frequency of occurrence of each character among members of the phena (see Table 2.8). Such a 'frequency matrix' can be held in the computer and is immensely valuable as the database describing the quantitative basis of the classification. The information can be used in several ways. First, the distinctive features of taxa can be recognized in these matrices and used for identification purposes (see Chapter 7). For example, the data in Table 2.8 show that phenon 5 can be identified as Gram-positive cocci that are oxidase negative but have a fermentative metabolism. This weighting of important features for identification purposes is permissible because it is *a posteriori* weighting. The characters have been shown to be of useful diagnostic value and to be effective in separating the taxa.

Table 2.8 Characteristics of phenetic groups, in terms of percentage positive responses

Characteristics	Phena				
	1	2	3	4	5
Gram-positive	0	0	0	100	100
Presence of rods	100	100	100	100	0
Presence of cocci	0	0	0	0	100
Presence of endospores	0	0	0	100	0
Fermentative metabolism	100	100	0	0	100
Oxidative metabolism	0	0	100	100	0
Catalase production	87	95	100	82	77
Oxidase production	71	83	75	23	0

A second useful application of the frequency matrix is in the areas of ecology and biotechnology. Again in Table 2.8, phenon 3 has been shown to be a good source of amylases by virtue of its members all hydrolysing starch, and phenon 2 is a potential source of proteases. Such organisms might be worth screening for enzymes with novel properties which may be of commercial value. But how to isolate more of these strains? Fortunately, the frequency matrix also indicates how this may be achieved. Phenon 3 strains are resistant to streptomycin and grow in 7% NaCl. A medium containing one or a combination of these compounds should be selective for phenon 3 bacteria and restrict the growth of most others. Similarly, phenon 2 bacteria could perhaps be isolated selectively by providing inositol as a sole carbon source. Large frequency matrices can be searched by computer programs to highlight combinations of factors for inclusion in selective media for the isolation of rare or novel microbes (see Chapter 8; Bull *et al.*, 1992).

2.8 CONCLUDING REMARKS

Numerical taxonomy developed rapidly in the 1970s, matured in the 1980s and is now the elder statesman of bacterial systematics. As a means of providing essential physiological and biochemical information about a group of unknown organisms it has much to recommend it; particularly as the data can be held in a computer and developed for identification or other uses. Perhaps the decline of numerical taxonomy in the 1990s merely reflects the fact that most 'important' groups of bacteria have now been characterized by this approach and few, poorly studied taxa remain. However, this is probably compounded by the attraction and glamour of the new approaches to systematics based on molecular biology. Nevertheless, numerical classification provides the basic procedures and statistical methods for many of the other disciplines and as such remains an important component of microbial systematics.

REFERENCES

Alderson, G. (1985). The application and relevance of nonhierarchic methods in bacterial taxonomy, in *Computer-Assisted Bacterial Systematics* (M. Goodfellow, D. Jones and F.G. Priest, Eds.), pp. 227–263, Academic Press, London.

Austin, B. and Colwell, R.R. (1977). Evaluation of some coefficients for use in numerical taxonomy of microorganisms, *International Journal of Systematic Bacteriology*, **27**, 204–210.

Bull, A.T., Goodfellow, M. and Slater, J.H. (1992). Biodiversity as a source of innovation in biotechnology, *Annual Review of Microbiology*, **46**, 219–252.

Clifford, H.T. and Stephenson, W. (1975). *An Introduction to Numerical Classification*, Academic Press, New York.

Dunn, G. and Everitt, B.S. (1982). *An Introduction to Mathematical Taxonomy*, Cambridge University Press, Cambridge.

Goodfellow, M., Austin, B. and Dickinson, C.H. (1976). Numerical taxonomy of some yellow-pigmented bacteria isolated from plants, *Journal of General Microbiology*, **97**, 219–233.

Goodfellow, M., Embley, T.M. and Austin, B. (1985). Numerical taxonomy and emended description of *Renibacterium salmoninarum*, *Journal of General Microbiology*, **131**, 2739–2752.

Logan, N.A. and Berkeley, R.C.W. (1981). Classification and identification of members of the genus *Bacillus* using API tests, in *The Aerobic Endospore-Forming Bacteria: Classification and Identification* (M. Goodfellow and R.C.W. Berkeley, Eds.), pp. 105–140. Academic Press, London.

Sackin, M.J. (1987). Computer programs for classification and identification, *Methods in Microbiology*, **19**, 459–494.

Sneath, P.H.A. and Johnson, R. (1972). The influence on numerical taxonomic similarities of errors in microbiological tests, *Journal of General Microbiology*, **72**, 377–392.

Sneath, P.H.A. and Sokal, R.R. (1973). *Numerical Taxonomy: The Principles and Practice of Numerical Classification*, W.H. Freeman, San Francisco.

CHAPTER 3

Chemosystematics and molecular biology I: Nucleic acid analyses

3.1 INTRODUCTION

With the development of fast and reliable analytical techniques in chemistry and molecular biology, the reliance on traditional micro-biological tests for gathering phenotypic data has decreased. Instead, a wealth of new information has been obtained through chemosystematics or chemotaxonomy, defined as the study of chemical variations in living organisms, and the use of these chemical characters for classi-fication and identification. This has broadened the database on which classifications are based. For some taxa, such as the actinomycetes, in which there are few useful physiological and morphological characters, chemosystematics, in particular, cell wall and lipid analyses, has been invaluable. For all bacteria, nucleic acid analyses are providing a sound taxonomic framework within which other approaches can be integrated. There have been suggestions that chemotaxonomic data are, in some sense more fundamental, or more closely a reflection of the genome than morphological traits, and will give more accurate classifications. Apart from nucleic acid sequences themselves, this does not seem to be the case, and biochemical, physiological and morphological classi-fications should, and do, resemble each other. In view of the connection between the two it could hardly be otherwise. Chemosystematics, therefore, provides us with a powerful supplement to the traditional approaches to classification and identification.

The levels within the cell at which chemotaxonomy operates and the most appropriate points within the taxonomic hierarchy at which these levels apply are illustrated in Table 3.1, from which it is clear that chemical characteristics may be used to establish relation-ships at all levels within the taxonomic hierarchy. Thus, DNA

50

Table 3.1 Chemosystematic analyses of the bacterial cell and the taxonomic level at which they are generally most useful

Cell component	Analysis	Taxonomic rank
Chromosomal DNA	Base composition (%G+C)	Genus
	DNA:DNA reassociation	Species
	DNA restriction patterns	
	rRNA restriction fragment length polymorphisms	Species and subspecies
Ribosomal RNA	Nucleotide sequence	Species, genus and above
	DNA:rRNA hybridization	
Proteins	Amino acid sequence	Genus and above
	Serological comparisons	Species and genus
	Electrophoretic patterns	
	Multilocus enzyme electrophoresis	Clones within species
Cell walls	Peptidoglycan structure	
	Polysaccharides	Species and genus
	Teichoic acids	
Membranes	Fatty acids	
	Polar lipids	
	Mycolic acids	Species and genus
	Isoprenoid quinones	
Metabolic products	Fatty acids	Species and genus
Complete cell	Pyrolysis – gas liquid chromatography	
	Pyrolysis – mass spectrometry	Species and subspecies

sequence comparisons and electrophoretic protein patterns are useful at the species and subspecies levels and, at the other extreme, rRNA analyses are being increasingly used to delineate major divisions among all living organisms. This fine specificity, and yet ability to embrace wide taxonomic distance, is a particularly valuable aspect of these approaches.

Chemosystematic analyses are phenetic in nature. Sequence data are often referred to as phylogenetic, but the data are phenetic. Relationships between organisms are being assessed on the present-day structure of those organisms, and the classifications obtained are phenetic, as are those derived from numerical analysis of the phenotype (see Chapter 2). In many instances, chemosystematic and sequence data have been processed using similarity coefficients and clustering or ordination algorithms to produce dendrograms and scatter diagrams analogous to those derived by traditional numerical taxonomy methods. Such phenetic classifications have often been considered to be phylogenetic, but are only a true reflection of cladogeny if evolution has been divergent and at constant rates. More recently, explicitly cladistic methods have been used to analyse sequence data from rRNA and protein-coding genes and the phylogenetic relationships so revealed are designed to reflect pathways of evolutionary descent. This phylogenetic approach to classification using nucleic acid and protein sequences and the methods used to construct phylogenies will be discussed in Chapter 5, but the analyses themselves and the phenetic (and phylogenetically relevant) classifications so derived will be described below.

One of the major drawbacks of chemosystematics is the dependence of the chemical composition of micro-organisms on the environment. Bacteria change their chemical composition substantially to accommodate environmental fluctuations. Thus, when comparing bacteria on the basis of some chemical component, it is important that the variation observed is a result of genetic differences and not due to an environmental effect. Cultures must, therefore, be grown under identical conditions and to the same stage of the batch culture growth cycle to ensure uniformity of environmental influence. This may be particularly difficult, sometimes impossible, if physiologically diverse organisms, such as thermophiles and psychrophiles or aerobes and microaerophilic taxa, are being compared.

Of the various chemical components used for taxonomy (Table 3.1), only chromosomal DNA and RNA are unaffected by growth conditions. The amounts of these molecules will fluctuate with growth rate, but the composition is invariant. Thus nucleic acids offer the only standard molecules by which the widest range of micro-organisms (and higher eukaryotes) can be compared and classified. Constitutively synthesized proteins are also very useful in this respect but individual proteins may not be distributed universally. The cell walls of micro-organisms vary widely in composition depending on the ionic complement of the medium. For example, phosphate limitation results in a total replacement of the phosphate-containing teichoic acids by teichuronic acid in the cell walls of some Gram-positive bacteria. The lipid composition of membranes is heavily influenced by temperature; at low temperatures, unsaturated fatty acids tend to dominate and are replaced by saturated fatty acids at high temperature. The composition of the complete cell will also vary according to the environmental conditions. It must, therefore, be remembered that, although chemosystematics has much to offer the bacterial taxonomist, in instances other than nucleic acid analyses, the cultural conditions must be carefully standardized and the possibility of environmental effects recognized.

In this chapter, we will examine nucleic acid analyses in the context of microbial systematics; the other chemical components of the cell will be discussed in Chapter 4.

3.2   CHROMOSOMAL DNA

The chromosome may be analysed at three levels. Firstly, the gross composition of the chromosome as the content of the four bases is of value. The duplex structure of DNA and hydrogen bonding between guanine (G) and cytosine (C), and between adenine (A) and thymine (T) bases, ensures that equivalent amounts of G+C and A+T are represented in the molecule. The content of the G+C (mol% G+C) in the molecule from bacteria varies within very broad limits from a minimum of about 25% (e.g. some mycoplasmas) to a maximum of approximately 75% (e.g. some streptomycetes). Beyond these limits, the genetic code would be so skewed that few sensible proteins could be constructed.

Secondly, an assessment of sequence homology between DNA molecules from two bacteria may be made by measuring the extent of renaturation of DNA molecules from the two sources in DNA 'reassociation' or 'pairing' experiments. An important aspect of DNA reassociation experiments is that the two molecules must be about 80% homologous for duplex molecules to form through base pairing. Less than 80% sequence homology and no hybrids can be formed. Therefore, the DNA reassociation assay estimates a difference of about 20% spread between 0 and 100%, where the latter is defined by the maximum reassociation obtained with the homologous DNA strands. Nevertheless, these crude estimates of DNA sequence homology are immensely valuable for classification and identification of closely related bacteria, such as strains within a species.

Finally, DNA sequencing has developed to such an extent that it is now possible to compare sequences of individual genes directly for several strains. This type of information is valuable in the phylogenetic context and for population genetic studies but the labour and time involved in generating the sequences lessens its applicability to general systematics, for the time being.

3.2.1   Methods for determining the base composition of DNA

Perhaps the most time-consuming aspect of all nucleic acid analyses is the preparation and purification of the material. Methods for the isolation of bacterial DNA are described in detail elsewhere (e.g. Owen and Pitcher, 1985). Essentially, cells are lysed with a detergent, such as sodium dodecyl sulphate. Gram-positive cells have first to be digested with lysozyme. Protein is removed by digestion with a non-specific protease (e.g. pronase) and chemical deproteinization using phenol or chloroform. RNA is removed with RNAase, and DNA can be selectively precipitated in the presence of ribooligonucleotides by isopropanol. DNA is concentrated and further purified by precipitation in ethanol, although carbohydrate contamination sometimes persists. This may be removed by hydroxylapatite chromatography or treatment with concanavalin A. It is important that the DNA is essentially free from protein, RNA and carbohydrate, because these impurities can interfere in subsequent assays for determining base composition or reassociation.

Direct estimation of the nucleotide composition of DNA involves hydrolysis followed by separation and quantitation of the products by chromatography. Although the introduction of high performance liquid chromatography (HPLC) has made such procedures more rapid and accurate, they remain less popular than the physicochemical approaches. These measure some physical parameter of the duplex molecule and relate this to mol% G+C using empirical formulae. A somewhat outdated but accurate approach is isopycnic equilibrium centrifugation in caesium chloride (CsCl) gradients. Centrifugation of a concentrated CsCl solution at high speed results in a density gradient. If a sample of DNA is incorporated into the CsCl solution, as the gradient is generated, so the DNA migrates to form a band at a point on the gradient that is equal to its own density (the isopycnic point). The buoyant density of the unknown can be calculated from the position of this band relative to the position of a standard DNA of known buoyant density. There is

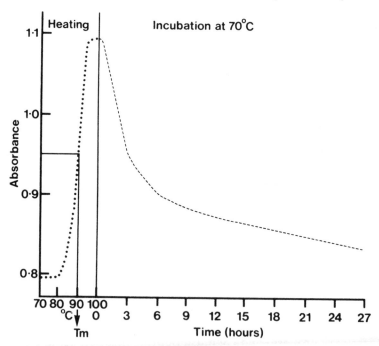

Fig. 3.1 Denaturation and reassociation of *Nocardia farcinica* DNA. Note the change in scale of abscissa from temperature to time (adapted from Bradley and Mordarski, 1976).

a direct relationship between buoyant density and mol% G+C, and thus the G+C content can be calculated (buoyant density = 1.66 + 0.098 (G+C)). This method is accurate and relatively insensitive to contamination of the DNA by protein or RNA, but the expense and complexity of the analytical ultracentrifuge have resulted in its declining popularity.

A second physical parameter of DNA related to base composition is the temperature at which the two strands separate. Thermal denaturation of duplex DNA is accompanied by an increase in absorbance at 260 nm (the absorbance maximum for DNA). The $A_{260 \text{ nm}}$ of a preparation of native DNA will increase by about 40% during denaturation – the hyperchromic shift (Fig. 3.1). The mid-point of the hyperchromic shift is referred to as the melting temperature (T_m) and is linearly related to mol% G+C. As with buoyant density, T_m can be converted to mol% G+C using an established empirical formula: G+C (%) = 2.44 T_m–169. Thermal denaturation is a popular method for the determination of base composition, since it is relatively insensitive to RNA or protein contamination of the DNA, and is rapid and inexpensive.

Both physicochemical procedures require comparison of an unknown against a standard or reference DNA. This is usually *Escherichia coli* DNA (51 mol% G+C) but, because different laboratories use different values for *E. coli* DNA, care must be taken in comparing results.

3.2.2 Taxonomic value and applications of DNA base composition

Bacteria differ widely in base composition, but the G+C content is constant for a given organism. Similarity in base composition between two organisms does not necessarily imply relatedness, e.g. the G+C ratio of *Spirochaeta halophila* and *Pseudomonas testosteroni* is 62% but these bacteria are distantly related. Since base composition does not take into account the linear sequence of bases in DNA, two organisms may have identical base compositions, but very few, if any, sequences and hence proteins in common. However, the converse is applicable. If two organisms possess DNA with widely different base composition, they will have few DNA sequences in common and are likely to be distantly related. Base composition is, therefore, a negative criterion; differences in base composition signify differences in nucleotide sequence in the DNA and hence dissimilar organisms.

It follows that DNA base composition provides a useful measure of genetic heterogeneity. If a genus contains species that differ widely in base composition, for example, the aerobic, endospore-forming bacteria of the genus *Bacillus* have DNA which varies from 32% to 67% G+C, it may be shown theoretically that organisms such as *B. cereus* (33% G+C) and *B. stearothermophilus* (53% G+C) can have few, if any, DNA sequences in common and should be allocated to different genera. Indeed from data gathered since 1960, it has been suggested that a

Table 3.2 Base compositions of some common bacterial genera

Genus (Gram negative)	Mol% G+C	Genus (Gram-positive)	Mol% G+C
Acetobacter	51–65	*Bacillus*	33–64
Aeromonas	57–63	*Clostridium*	22–55
Cytophaga	29–45	*Corynebacterium*	51–63
Enterobacter	52–60	*Lactobacillus*	32–52
Escherichia	48–52	*Micrococcus*	66–75
Haemophilus	37–44	*Mycobacterium*	62–70
Klebsiella	53–58	*Staphylococcus*	30–37
Pseudomonas	58–70	*Streptococcus*	34–46
Rhodospirillum	60–67	*Streptomyces*	69–77

useful guideline might be that the maximum genetic variation permissible within a genus should be that represented by 10–12% G+C. Thus, genera such as *Bacillus* (33–64%), *Lactobacillus* (32–52%), *Flavobacterium* (31–68%) and *Haemophilus* (37–55%) probably each require division into several genera. Most bacterial genera, however, have comparatively narrow ranges of G+C values which conform to this overall scheme (see Table 3.2).

Base composition of DNA is also useful at the species level. It has been suggested that members of a species should differ by no more than 5% G+C; deviation beyond this limit being indicative of excessive genetic dissimilarity. Differences in base composition within this 5% variation can be valuable, but it must be remembered that methods for determining mol% G+C are seldom sufficiently accurate to distinguish a variation of less than 2%. However, *B. megaterium* strains have been shown to comprise two groups from estimates of base composition; group 1, 37–38% and group 2, 40–41%. Subsequent numerical taxonomy analyses revealed that two phenetically different species-ranked taxa were represented by these groups of strains; *B. megaterium sensu stricto* and *B. flexus*. Thus, DNA base composition is a useful taxonomic tool to the extent that it is required information for the description of any new bacterial taxa.

It is tempting to speculate on the reason for the heterogeneity of base composition among micro-organisms which covers some 50% for bacteria, 30–70% for yeasts and fungi, 22–66% for protozoa, and 36–68% for algae. Given a typical bacterium with a G+C content of 42% and average protein composition, it may be shown that the redundancy of the genetic code would allow limits of 31% and 67% G+C to code for the proteins of the organism. However, there is slight variation in the protein composition of code-limit species, and, allowing for this, it can be calculated that their theoretical extremes would be 28% and 72%. These are very close to the observed extremes. Presumably, there is some selection pressure to extremes of G+C, but the simple schemes of high ultraviolet environments selecting for high G+C chromosomes (lack of A+T for the induction of thymine dimers) or high temperature environments selecting for more thermostable (high G+C) chromosomes are not consistent with the observed patterns. Alternatively, it may be a random process.

3.2.3 Methods for determining gross DNA sequence homology

Assessment of base sequence similarity between DNA from two organisms may be readily achieved by DNA reassociation experiments, in which DNA is rendered into single strands by thermal or alkali denaturation and subsequently allowed to anneal in the presence of a second denatured DNA molecule. If the nucleotide sequences of the two DNA samples are largely homologous, hybrid duplexes will be formed by base pairing. If there are few sequences in common, there will be negligible hybrid formation (Figs. 3.2 and 3.3). This technique, therefore, provides a quantitative estimate of DNA sequence homology between two organisms.

There are essentially two approaches to the determination of DNA base sequence complementarity; free solution assays (Fig. 3.2) and immobilized DNA assays (Fig. 3.3). The key feature is that a large excess (at least 1000-fold) of single stranded (ss)DNA is sheared to a constant molecular weight and incubated with radioactively labelled (usually ^{32}P) sheared ssDNA from a reference strain to allow renaturation to occur. Two controls, comprising (1) the homologous reaction and (2) labelled DNA with salmon sperm DNA (which has no homology with bacterial DNA), are included. After renaturation, three types of molecule will be present; (1) ssDNA from the reference strain which has not hybridized to the excess of unlabelled DNA and cannot hybridize to itself because of its low concentration, (2) dsDNA which represents renatured DNA from the large excess of unlabelled DNA and (3) hybrids where the labelled reference DNA has renatured with the excess of unlabelled DNA. It is necessary to separate the double stranded (ds)DNA from the remaining ssDNA. This may be achieved by treating the mixture with a nuclease from *Aspergillus oryzae* (SI nuclease) that hydrolyses ssDNA but not dsDNA. After digestion, intact dsDNA is precipitated with trichloracetic acid, filtered and the radioactivity in the precipitate counted (Fig. 3.2). Percentage homology values are obtained by dividing the counts per minute in the heterologous, SI-resistant DNA, by the activity of the homologous reaction, and multiplying by 100. The non-specific, background reaction is provided by the counts in the salmon DNA control.

An alternative scheme for separating the ds from ssDNA is hydroxylapatite chromatography. In low molarity, phosphate buffers,

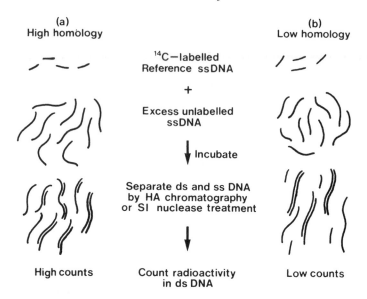

(a)
High homology

(b)
Low homology

^{14}C—labelled
Reference ssDNA

+

Excess unlabelled
ssDNA

↓ Incubate

Separate ds and ss DNA
by HA chromatography
or SI nuclease treatment

High counts   Count radioactivity   Low counts
              in ds DNA

Fig. 3.2 Schematic representations of free solution DNA reassociation assays in which labelled reference DNA is incubated with an excess (at least 1000-fold) of unlabelled DNA from a second organism.

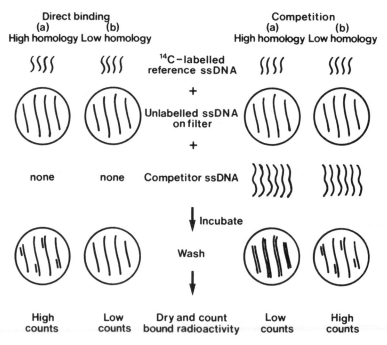

Direct binding   Competition
(a)   **(b)**   **(a)**   **(b)**
High homology Low homology   **High homology Low homology**

^{14}C—labelled
reference ssDNA

+

Unlabelled ssDNA
on filter

+

none   none   Competitor ssDNA

↓ Incubate

Wash

↓

High   Low   Dry and count   Low   High
counts   counts   bound radioactivity counts   counts

Fig. 3.3 Schematic representation of immobilized DNA reassociation assays performed either by direct binding or with competitor DNA (see text for details).

only dsDNA will bind to hydroxylapatite, ssDNA is not adsorbed. The dsDNA can then be eluted with high molarity buffer, precipitated and counted in a scintillation counter. As before, percentage homology is the amount of label incorporated in the heterologous reaction divided by that in the homologous reaction, and multiplied by 100.

Free solution reassociation assays can be conducted without recourse to labelling DNA. The reassociation of ssDNA is accompanied by a reduction in $A_{260\ nm}$ (Fig. 3.1). The initial rates of reassociation of an equimolar mixture of DNAs from two organisms are compared with the initial rates of the two homologous reactions. Non-homologous DNAs will reassociate more slowly (if at all) than the homologous molecules, and the relative rates may be used to determine percentage of homology from established formulae. This method has the advantage of being rapid; labelled DNA is not required and comparisons may be made between any pair of organisms rather than between reference and test strains. Moreover, with care, it is as accurate as other procedures (see Owen and Pitcher, 1985; Johnson, 1991).

Immobilized DNA reassociation assays use ssDNA bound to a solid support to effect the separation of renatured dsDNA from labelled ssDNA (Fig. 3.3). Single stranded DNA will bind to nitrocellulose and purpose-made nylon filters in such a way that the molecules retain the ability to form a duplex through hydrogen bonding of complementary sequences. In the commonly used direct assay, filters, containing immo-bilized ssDNA from different organisms, are incubated with labelled reference ssDNA to allow renaturation to occur. The membranes are subsequently washed, dried, and counted in a scintillation counter. The extent of homology is calculated as for the free solution assays described above, and high counts bound to the filter indicate high homology.

This method has been made more popular by the introduction of blotting manifolds. These perspex filter holders allow up to 96 denatured DNA samples to be blotted onto a single nitrocellulose or nylon mem-brane in a distinctive array. The filter is then hybridized to labelled probe DNA, washed, dried and an autoradiograph may be prepared, such as that shown in Fig. 3.4. The filter can then be cut into slices and individual reactions counted in a scintillation counter to provide quanti-tative estimates of sequence homology. This method, which is routinely used for qualitative work in molecular biology, is less precise than the free solution assays (Grimont, 1988) but with careful standardization of the amount of DNA loaded onto the filters and triplicate determina-tions, sufficiently reproducible results can be obtained for routine purposes.

Competition reassociation is a variation of the direct immobilized assay and has the advantage that only reference DNA is bound to the membranes. This improves the accuracy of the method insofar as there

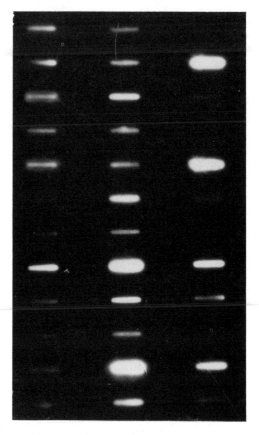

Fig 3.4 Autoradiogram of a DNA reassociation experiment carried out using a 'slot-blot' apparatus. The homologous reaction is shown in the second row down of the right-hand lane, and the heterologous control immediately above the homologous reaction. The filter is cut into 36 individual reactions and each is counted in a scintillation counter to provide estimates of DNA sequence homology (see text for details).

is less variability in the amount of DNA bound to filters. In competition experiments, a small amount of labelled reference ssDNA competes for reassociation sites on the homologous reference ssDNA immobilized on the filter with an excess of heterologous competitor ssDNA. If the competitor DNA is complementary to the reference DNA, it will preferentially reassociate with the bound DNA because of its higher concentration. Conversely, if there is little or no homology between the competitor and immobilized DNA molecules, the labelled reference DNA will reassociate with the membrane-bound DNA. Thus, low counts on the washed dried membrane indicate high homology, and high counts represent low homology between the reference and heterologous competitor DNAs (Fig. 3.3). Methods for DNA reassociation have been fully described by Owen and Pitcher (1985) and Johnson (1991).

All DNA reassociation assays must be carefully standardized if they are to give reproducible results since the extent and specificity of reassociation is heavily influenced by external conditions and the physical state of the DNA. There are essentially five factors that affect DNA reassociation assays (Johnson, 1991):

1. DNA reassociation is influenced by the size of the DNA fragments; the larger the fragment the greater the rate of reassociation in free solution, but, as the viscosity of the solution increases, the rate of the reaction decreases. Moreover, very small fragments (less than about 15 nucleotides in length) show little specificity during renaturation; a compromise is therefore required. It is usual to reduce the size of the fragments by physical shear using sonication or pressure drop in a French press to a uniform molecular weight of $1–5 \times 10^5$ daltons, representing about 200–1000 base pairs.

2. The rate and extent of renaturation increases as the ionic strength of the incubation buffer is increased. DNA will not renature in the absence of salts owing to the repulsion between negatively charged phosphate groups in the molecules. Duplexes formed at high ionic strength are less temperature stable than those formed at lower ionic strength, indicating that a loss of specificity in pairing occurs at the higher ionic strength. It is, therefore, important to use standard buffers for these experiments. These buffers are usually based on standard saline citrate (SSC) which is 0.15 M NaCl and 0.015 M trisodium citrate (pH 7.0) in which the Na ions repress the repelling effect by shielding the anionic phosphate groups and the citrate is a chelating agent to inhibit nucleases by binding divalent cations.

3. Purity of the DNA preparation is important; contamination with RNA or carbohydrates can interfere in the extent of reassociation. It is also important to ensure that the DNA samples are completely denatured into single strands by thoroughly boiling the DNA and cooling it rapidly on ice.

4. DNA concentration and time of incubation are critical features in reassociation assays. Britten and Kohne (1966) introduced the concept of 'cot' and showed that duplex formation is a function of the initial DNA concentration (C_0) and the time of incubation (t). DNA reassociation is a second order reaction in which the change in substrate concentration with respect to time (dc/dt) is proportional to the square of the concentration:

$$-dc/dt = k^1 c^2$$

The integral form of this equation can be rearranged as shown below:

$$C/C_0 = 1/(1 + kC_0 t)$$

The concentration of ssDNA after t seconds is C and the reassociation rate constant is k. The ratio C/C_0 is the fraction of DNA that is denatured at any one time and can be plotted against C_0t values to provide a 'C_0t' curve as shown in Fig. 3.5. There are two important

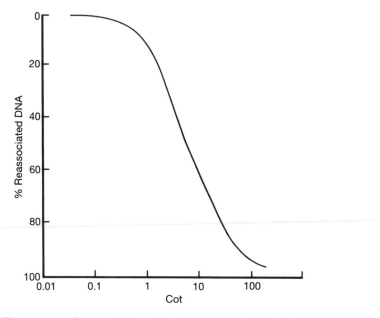

Fig. 3.5   Time course of reassociation of *E. coli* DNA. Labelled and unlabelled DNA fragments from *E. coli* were incubated to various *Cot* values and the amount of reassociation determined (see text for details; adapted from Brenner, 1970).

features of this curve. The value for 50% reassociation is designated $C_0t_{1/2}$ at which point $C_0t = 1/k$. The value of $C_0t_{1/2}$ reflects genome size or complexity and will be reached more rapidly when smaller (less complex) DNA molecules are being examined. This can therefore be used to estimate the molecular weights of bacterial chromosomes. At high C_0t values, the amount of DNA reassociation reaches 100% and continuation of incubation will not increase the amount of DNA reassociation. Most hybridization reactions are continued for no more than 100 C_0t·s. In free solution assays, it is impossible to distinguish between renatured labelled DNA and the hybrid comprising labelled and non-labelled DNA. By using a 4000-fold excess of unlabelled DNA (i.e. 0.1 µg ml^{-1} labelled DNA and 400 µg ml^{-1} unlabelled DNA) incubated for 21 h, the labelled DNA receives a C_0t of about 100 and the labelled DNA a C_0t of 0.025, which would result in about 1% reassociation of labelled DNA. For immobilized DNA assays, a lower ratio (100:1) is usually sufficient to ensure sufficient sites on the filter for the labelled DNA to reassociate, and the homologous labelled DNA is not a serious problem.

5. The temperature, at which the reassociation mixture is incubated, is critically important. The optimal renaturation rate occurs at about 25°C below the T_m of the DNA mixture. Below this temperature, non-specific hybrids are formed and distantly related sequences reassociate. Above this temperature, the pairing is more stringent, to the extent that the reaction is reduced considerably and only closely related sequences form stable duplexes. It is usual to work at 25°C below T_m for general studies, but, if closely related bacteria are being examined, the more stringent 10–15°C below T_m is used.

By careful attention to these parameters, DNA reassociation studies are accurate and repeatable and, despite the variety of procedures available, congruent results are obtained.

3.2.4 Taxonomic value and applications of DNA reassociation

DNA reassociation is recommended for the evaluation of phenetic relationships for several reasons:

1. The DNA composition of a cell is invariant regardless of the growth conditions, although the content will vary with growth rate. Thus, the stage of growth (lag, exponential, or stationary phase, even sporulated cells) or nature of the medium will have no effect on the reproducibility of the results. As a consequence, classifications based on DNA reassociation tend to be more stable than those derived from other procedures.

2. A second feature of DNA reassociation, that contributes to the stability of the classifications it provides, is that the estimation of relatedness between organisms is based on the complete genotype. Numerical phenetics compares organisms on the basis of a fraction of the genotype (perhaps 5–20%) and this is valid so long as a representative sampling of the genome is achieved. This is attempted by analysing a heterogeneous battery of characters and, in general, the aim of representative sampling is achieved, since classifications from DNA reassociation and numerical phenetics are largely congruous. Data for % similarity and DNA reassociation for some enterobacteria are shown in Fig. 3.6. The relationship between numerical phenetics and % DNA pairing is not a straight line since there is very little sequence homology at phenetic similarities below about 50% S_{SM}. The presence of apparent phenetic similarity at low DNA sequence homology probably results from negative matches (absence of a gene(s)) which are included by the S_{SM} coefficient (see Chapter 2) as a measure of similarity and yet would not be recognized as shared DNA sequences. The S_{SM} coefficient may therefore give

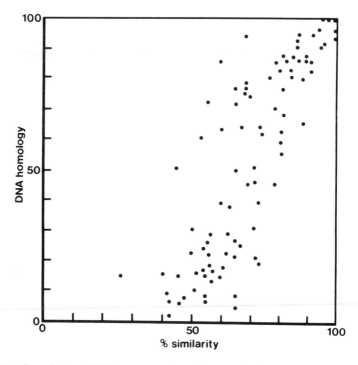

Fig. 3.6 Correlation of DNA sequence homology with phenetic similarity (adapted from Staley and Colwell, 1973).

rise to erroneously high measures of similarity. Even with the S_J coefficient, which ignores negative correlations, there is a similar deviation from linearity in the relationship between numerical taxonomy and DNA sequence homology at low similarity levels. This is probably due to the shared presence of an attribute (e.g. gas production from glucose, or resistance to an antibiotic) which would be considered a match by the S_J but may be due to entirely different enzymes and consequently different DNA sequences. Moreover, the same protein may be encoded by completely different DNA sequences because of the redundancy of the genetic code. The reverse situation, in which homology values are higher than numerical phenetic similarity values, at high similarity levels is not so pronounced and probably results among highly related bacteria from slightly mismatched duplexes formed under optimal conditions being estimated as 100% reassociated. Areas of localized mismatched bases, or deletion mutations, are not detected by the relatively crude methods used for processing large numbers of reassociation determinations, and yet could be responsible for lack of a particular attribute and estimated as a difference in phenotype by numerical taxonomy. It is, however, encouraging that there is this general congruence between DNA reassociation and numerical phenetics that lends credibility to both methods.

3. DNA reassociation analyses may be refined by estimating thermal stabilities of reassociated duplexes. Hybrid DNA molecules generally have a lower T_m than parental homologous molecules resulting from mismatched sequences; each 0.7°C drop in T_m represents 1% mismatched bases. By measuring the reduction in T_m (ΔT_m) for a hybrid duplex, it is possible to estimate the extent of sequence divergence. Thus, a hybrid duplex formed at 30°C below T_m between *E. coli* and *Serratia marcescens*, would reassociate about 25%. The ΔT_m would be 13–14°C, indicating about 20% sequence divergence in the reassociated sequences. Conversely, DNA from *E. coli* and *Shigella dysenteriae* would reassociate 81–90% with a ΔT_m of 1–6°C, representing only some 4% mismatched sequences. This provides an estimate of affinity between very closely related bacteria. If *E. coli* were the common ancestor, it can be argued that *Ser. marcescens* diverged earlier in evolutionary time than *Sh. dysenteriae*. An alternative to ΔT_m is to measure the % reassociation at optimal and stringent temperatures, and divide the latter figure by the former. This ratio, the divergence index or thermal binding index, approaches 1.0 in closely related bacteria that have few sequence dissimilarities, and 0.00 in unrelated bacteria. It is less laborious than ΔT_m determinations.

4. DNA homology is providing a unifying concept of the bacterial species. A bacterial species is usually defined as a group of strains that share many features, and differs considerably from other groups (see Chapter 6). This definition is subjective, and, it is not surprising that some species are far more heterogeneous than others. A more objective and rigorous definition of species is desirable, and DNA homology studies are indicating how this might be achieved. By examining DNA reassociation data gathered since the techniques were introduced in the early 1970s, it has become apparent that organisms within well defined species have extensive DNA sequence homology. Moreover, organisms representing different species have very few sequences in common. The lower allowable limit of sequence homology within a species has been a controversial topic, but a widely agreed definition is 70% relatedness measured under optimal conditions. Thus, strains within a species should show more than 70% relatedness and less than 5% divergence as measured by ΔT_m. This is proving to be a definition of great practical importance since it attempts to provide a universal definition of the prokaryotic 'genomic' species.

5. Finally, the application of DNA probe technology is an extension of the early DNA reassociation experiments and is providing rapid

Nucleic acid analyses

and highly specific procedures for detection and identification of bacteria (see Chapter 7).

There are numerous examples of how DNA pairing studies have advanced bacterial systematics. One is provided by *Bacillus sphaericus* which is of interest because some strains of this species are pathogenic to mosquito larvae and can be used for the biological control of these vectors of disease. This bacterium is unusual among the bacilli by virtue of its differentiation into a spherical endospore; endospores are usually oval. Moreover, the organisms have a strictly oxidative metabolism for which acetate and other organic acids are the preferred carbon and energy sources. As a result, these bacteria are unreactive in most of the usual classification tests which depend on carbohydrate metabolism and show the same, largely negative phenotypes. All round-spore-forming bacteria were traditionally called *B. sphaericus*.

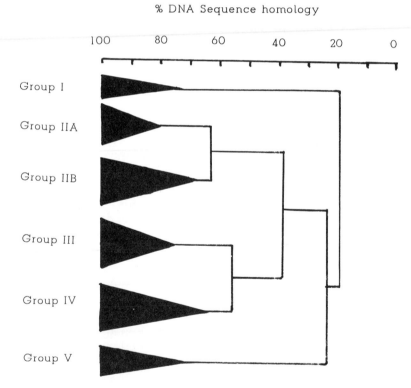

Fig. 3.7 A dendrogram showing the relationships between some round-spore-forming bacilli. Each block of the dendrogram represents numerous strains of high sequence similarity. *Bacillus sphaericus sensu stricto* strains were recovered in group I, and the insect pathogenic strains in group IIA. Each DNA homology group represents a species-ranked taxon (adapted from Krych *et al.*, 1980).

Extensive DNA reassociation studies showed that at least five homology groups or 'species' could be distinguished among '*B. sphaericus*' strains and the mosquito pathogens were all allocated to group II which was separate from group I, *B. sphaericus sensu stricto* (Fig. 3.7). Group II could be further divided into subgroups between which there was 60–70% sequence homology with a ΔT_m of about 5°C. All mosquito pathogenic strains were recovered in group IIA. The realization that the mosquito pathogens comprised a separate species from *B. sphaericus sensu stricto* has enabled the formulation of DNA probes for the detection and identification of these important bacteria in the environment. Furthermore, a numerical taxonomic study based on carbon source utilization rather than sugar fermentation tests recovered all the DNA homology groups as distinct phena and supported the division of the taxon into at least five species (Priest, 1992).

DNA reassociation techniques also have their limitations. Because of the amount of work involved, full similarity (reassociation) matrices with estimates of DNA homology between each and every strain are seldom produced. Instead, comparisons are made between judiciously chosen reference strains and a variety of test strains. Generally, this approach is reliable, but the reference strains must represent the taxa under study. If they do not, distortion of the taxonomic structure may occur. It is therefore, prudent to use DNA reassociation in conjunction with a second approach, such as traditional numerical taxonomy. The latter would provide boundaries for taxa, which can then be validated using DNA reassociation based on sequence homology with centrotype strains as references. Numerical taxonomy would also provide the diagnostic features for the construction of identification schemes. This combination is a powerful and reliable approach to bacterial classification and identification.

3.2.5 DNA restriction patterns and RFLP analysis

Restriction enzymes cleave DNA molecules at specific sites, generally four or six base pairs (bp) in length. Such sites occur on average every 256 or 4096 base pairs respectively. Digestion of chromosomal DNA with a 6 bp cutting enzyme therefore produces a range of fragments varying in size from less than 100 bp to 10 or more kilobase pairs (kbp). These fragments can be separated by electrophoresis in an agarose gel, stained with ethidium bromide and the banding pattern visualized in ultraviolet light. The conservation of sequence homology between strains of the same species extends to conservation of restriction sites, such that the banding patterns are consistent (with minor variations). The pattern, which represents many overlapping and duplicated bands can be recorded using a densitometer and stored in a computer; alternatively the patterns can be compared visually (Fig. 3.8). This procedure

I ND IIA IIA IIB IIB IIB III ND IV IV V V

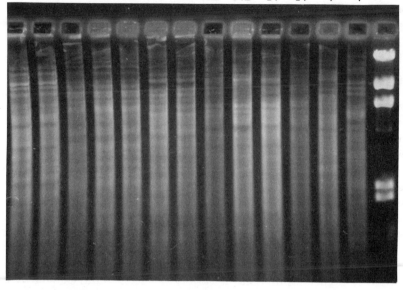

a

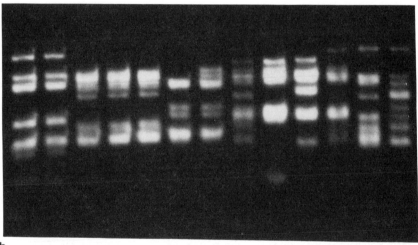

b

Fig. 3.8 Ribosomal RNA RFLP analysis of *Bacillus sphaericus* strains. (a) Chromosomal DNA from 13 *B. sphaericus* strains digested with *Hind*III and electrophoresed in agarose. The DNA homology groups of the strains are indicated (ND, not determined). Note that lane 2 contains a strain of DNA homology group I, and lane 5 contains a strain previously mislabelled as IIB. (b) The gel in (a), blotted and probed with a ^{32}P-labelled fragment of a cloned 16S rRNA gene from *E. coli*. Note that the RFLPs clearly distinguish groups I, IIA and IIB although groups III, IV and V are not so readily identifiable. It is now clear that the strain in lane 5 should be assigned to group IIA (from De Muro *et al.*, 1992, with permission).

therefore provides a simple alternative to DNA reassociation studies for allocation of strains to a species (Grimont and Grimont, 1991).

An extension of this procedure is to use restriction enzymes that cut the chromosomal DNA at rare sites. For example, *Sma*I, which recognizes the sequence 5'-CCCGGG-3' should cut low G+C DNA relatively infrequently. Restriction enzymes that bind to 8 bp recognition sites such as *Not*I or *Fse*I similarly produce a few, large molecular weight fragments from chromosomal DNA which can be resolved using special electrophoresis techniques such as field inversion gel electrophoresis. In this approach, the DNA fragments are completely resolved (Fig. 3.9) and the bands can be compared and related to nucleotide sequence divergence as proposed by Nei and Lei (1979).

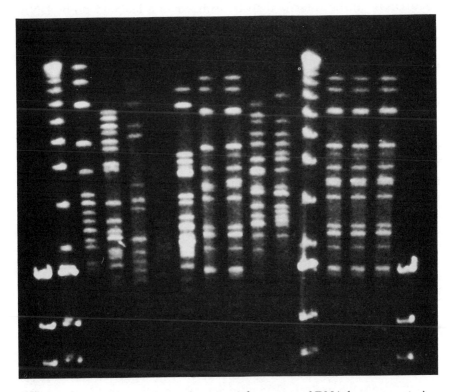

Fig. 3.9 Pulsed field gel electrophoresis of chromosomal DNA from some strains of *Neisseria* after digestion with restriction endonuclease *Nhe*I. The lanes from left to right contain: 1, λ DNA cut with *Hind*III (23.1, 9.4, 6.6 kb); 2, λ *Hind*III digest markers with λ concatemers (monomer size is 48 kb); 3, *N. gonorrhoeae*; 4, *N. flavescens*; 5, *N. lactamica*; 6, blank; 7, *N. polysaccharae*; 8, *N. meningitidis* serogroup B; 9, *N. meningitidis* serogroup B; 10, *N. meningitidis* serogroup A; 11, *N. meningitidis* serogroup A; 12, molecular weight markers; 13, 14 and 15, additional *N. meningitidis* strains (Figure courtesy of M. Maiden).

Restriction patterns can be simplified by highlighting certain bands by hybridization to a specific probe DNA and comparing the sizes and distribution of fewer bands, generally 5–20. Strains can then be compared on the basis of these restriction fragment length polymorphisms (RFLPs). To achieve the hybridization, the DNA is cut with a restriction endonuclease and electrophoresed through agarose. As with the above technique, it is essential that the DNA is cut to completion otherwise multiple banding patterns from composite fragments will be revealed and will confuse the RFLP pattern. The DNA can be visualized after electrophoresis to ensure that complete digestion has occurred. The DNA is then denatured by alkali treatment and transferred to a nitrocellulose or nylon membrane by capillary action (Southern transfer; see Ludwig, 1991 or any molecular biology methods text). The immobilized DNA on the filter is then hybridized to a labelled probe DNA (either radioactively or chemically labelled) and the banding patterns visualized.

The most common general probe is based on the 16S rRNA gene. The high sequence conservation between rRNA genes means that the 16S rRNA cistron from *E. coli* can be used as a multi-purpose probe for all bacteria although sometimes *B. subtilis* or *Staphylococcus aureus* rRNA is used for Gram-positive bacteria. The nature of the probe may vary. Originally, purified 16S rRNA was end-labelled with ^{32}P and used to hybridize with digested chromosomal DNA. An alternative was to use the cloned *E. coli* rRNA genes. The cloned fragment was excised from the plasmid vector, purified by agarose gel electrophoresis and labelled using nick-translation or random-primed labelling reactions (see Stahl and Amann, 1991 or any molecular biology text for details). The method of choice is, however, based on the polymerase chain reaction (PCR; see Chapter 5). Primers to the limits of the 16S rRNA gene are used with a chromosomal DNA template and a labelled deoxyribonucleotide is used in the reaction mix. In this way large amounts of 16S rDNA are generated with an incorporated label.

The target is the chromosomal DNA which has been digested and separated in an agarose gel before blotting onto the membrane. Bacteria contain between 1 (slow-growing mycobacteria) and 11 (*B. subtilis*) rRNA operons each comprising genes for 16S, 23S and 5S RNA separated by spacer DNA. The spacer regions may vary in sequence; far more so than the genes themselves. Depending on whether the restriction enzyme sites lie within or between the genes, reasonably complex RFLPs emerge, termed ribotypes, comprising 5–12 bands or even more (Fig. 3.8(b)). The banding patterns will differ depending on whether the rRNA itself or a PCR copy of the gene is used as probe compared with the cloned gene probe which may also include some spacer DNA.

Some specific probes have also been used to generate RFLPs. The term specific is used here to denote a gene which only occurs in a certain group of organisms as opposed to the universal rRNA gene probe. Cloned fragments of the cholera toxin gene have been used to distinguish between *Vibrio cholerae* isolates and a cloned fragment from the exotoxin gene has been used to type *Pseudomonas aeruginosa* (Grimont and Grimont, 1991).

3.2.6 Applications of DNA restriction patterns and RFLPs

The DNA sequence homology studies presented in Fig. 3.7 for *B. sphaericus* strains showed that the mosquito pathogenic strains allocated to homology group II were highly related to some non-pathogenic types. Indeed, group II was split into pathogens (group IIA) and non-pathogens (group IIB) related at 60–70% sequence homology. We used RFLPs to determine if these two subgroups were indeed different. The chromosomal DNA restriction patterns clearly distinguished the two types (Fig. 3.8(a)) and when probed with the cloned 16S rRNA gene

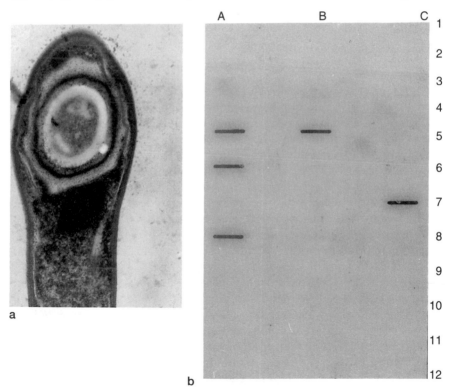

Fig. 3.10 (a) An electron micrograph of a thin section of *Bacillus sphaericus* showing the endospore and toxic crystal protein. (b) Hybridization of total chromosomal DNA to a labelled probe representing the toxin genes of a mosquito-pathogenic strain of *Bacillus sphaericus*. Only wells A5, B5, A6, C7 and A8 contain DNA from toxin-producing strains of *B. sphaericus* (Electron micrography courtesy of A. Yousten).

from *E. coli* two RFLPs were revealed, each specific for the DNA homology subgroups (Fig. 3.8(b)). This led us to suggest that the two subgroups represent different subspecies of the same species and that all mosquito pathogens represented a distinct taxon. Moreover, within group IIA, some strains are much more toxic than others. These high toxicity strains synthesize a 'crystal protein', which, when ingested by larvae is lethal (Fig. 3.10(a)). Low toxicity strains do not synthesize this crystal and are only mildly toxic to mosquito larvae. We have therefore used a specific probe, the cloned crystal protein genes, to identify highly toxic strains recovered from soil and insect samples. Only strains which have the capacity to produce a crystal protein hybridize to this gene probe (Fig. 3.10(b)). Using an RFLP approach and Southern blots, variation in crystal protein gene structure can also be detected and strains with novel toxin gene structures (and perhaps novel toxicity) identified. This method has also been valuable for distinguishing taxa within numerous Gram-positive and Gram-negative genera including; *Haemophilus, Providencia, Legionella, Campylobacter* and *Lactococcus* (see Rodrigues *et al.*, 1991 for references to original literature).

3.2.7   DNA sequencing

We feel that some mention should be made of DNA sequencing in a book devoted to microbial systematics because our classifications are ultimately based on the information in the chromosome, which is of course encoded in the sequence of nucleotides. Although chromosomes of few organisms (mainly viruses) have been fully sequenced, a complete chromosome from yeast has now been sequenced. The programme for generating the full sequence of *B. subtilis* is underway and sequence data for individual genes are being generated at an unprecedented rate. Some of the major target genes are the rRNA genes (see section 3.3) and other genes with conserved functions, such as those for elongation factor Tu that plays a central role in protein synthesis, ribosomal protein RplX and the beta-subunit of ATPase.

Virtually all DNA sequencing uses the dideoxy chain termination procedure that is described in all biochemistry and molecular biology books (e.g. Brown, 1992). This procedure is now being automated; in one approach different coloured fluorescent labels are used for the four nucleotides. The migration of the bands in the electrophoresis gel is detected by a laser and the sequence is compiled by computer. With the advancement of the 'Human Genome Project', automated sequencing of DNA is sure to become the norm and vast amounts of chromosomal sequence information will become available. But what will the systematist do with this wealth of data?

There are three major applications for comparative nucleic acid sequencing in systematics; all have a phylogenetic leaning and will be discussed more fully in Chapter 5. (1) Molecular phylogenies of particular genes or gene families can be constructed to trace the evolution of these genes and the evidence for lateral gene transfer. A study of beta-lactamases which shows that the gene for this antibiotic resistance enzyme has transferred throughout the bacterial kingdom is described in Chapter 5. (2) Organismal genealogies within species can be traced. This population genetics approach allows studies of individual relatedness, geographic variation and clonal structure of populations of strains within a species. It complements the multilocus enzyme electrophoresis approach (see Chapter 4) by providing more detailed information than the study of isozymes, but for fewer strains, because of the technical demands of generating the data. Sharp *et al.* (1992) have used sequence comparisons of the *rplx* (a ribosomal protein gene) locus of *B. subtilis* strains to demonstrate that the gene classification differed from the strain classification. This suggests that inter-strain recombination is common among strains of this species (see Chapter 4). (3) Species phylogenies can be constructed from the gene sequences as in the use of 16S rRNA sequences for cladistic classifications. This is described more fully below and in Chapter 5.

3.3 ANALYSIS OF RNA

There are three categories of RNA in prokaryotes; the short-lived messenger (m)RNA responsible for transmitting information from the chromosome to the ribosome, the stable form, i.e. transfer (t)RNA, which decodes the message, and ribosomal (r)RNA involved in the structure of the ribosome and the reading of the message. Analysis of RNA for taxonomic purposes focuses on the three rRNAs; the 5S, 16S and 23S molecules. These molecules are valuable as indicators of relatedness for the following reasons:

1. The rRNAs are essential elements in protein synthesis and are, therefore, present in all living organisms (with the notable exception of viruses).
2. Because of the conserved functions of these molecules they have changed very little during evolution. Thus, rRNAs from even the most taxonomically distant organisms, that share virtually no DNA sequence homology, will have rRNA sequences in common, and, therefore, relatedness can be assessed. Ribosomal RNA is probably unique amongst macromolecules in this respect.
3. Some segments of rRNA evolve more rapidly than others and sequence variation occurs between closely related organisms allowing comparisons to be made at the species level.
4. Phylogenetic lines of descent may be inferred from rRNA sequences (see Chapter 5).

Methods for comparing rRNA gene sequences began with rRNA:DNA hybridization in which the labelled rRNA from a reference strain is hybridized with chromosomal DNA from another bacterium. A limitation of this method is that distantly related organisms cannot be easily compared. Before DNA sequencing was a standard laboratory procedure, catalogues of rRNA molecules were generated in which enzymes were used to reduce the molecule to oligonucleotides which were then separated and sequenced. An average catalogue consisted of about 80 fragments (7–20 nucleotides in length) covering 35–45% of the complete sequence. More recently (Lane *et al.*, 1985), methods for the rapid sequencing of the complete larger rRNA molecules or their genes have been developed. Nevertheless, the period between the rise and fall of the 16S rRNA cataloguing technique was one of spectacular advance which culminated in the recognition of the archaea or archaebacteria as a coherent and independent phyletic lineage.

3.3.1   Sequence analysis of 5S rRNA

Prokaryotic 5S rRNAs may be made up of 120 nucleotides (from Gram-negative bacteria) or 116 (sometimes 117) nucleotides, as found in Gram-positive organisms. Sequences may be determined routinely for these relatively small molecules. Direct sequence analysis uses methods similar to the Maxam–Gilbert procedure developed for DNA sequencing. After radioactively labelling an end of the molecule, it is partially digested with base specific enzymes or reagents and the products are separated in four 'base specific' lanes by polyacrylamide gel electrophoresis (PAGE). This generates a ladder from which the base sequence can be read. To date, some several hundred 5S rRNA sequences have been determined.

It is necessary to align the sequences for comparisons to be made. The 120 and 116 nucleotide molecules can be aligned by inserting two residue gaps into the 5′ and 3′ ends of the 116 nucleotide molecule. Alignment with eukaryotic (120 nucleotide) molecules is more complex, and involves estimating the 'best match alignment' with minimal gap insertions (see Chapter 5). Simple comparisons of the molecules in terms of sequence homology can be clustered to provide a dendrogram. Alternatively, more complicated comparisons can be made that attempt to account for mutation and back mutation rates in order to provide a more accurate phylogenetic reconstruction (see Chapter 5).

The main criticism levelled at the 5S rRNA molecule as an indicator of relatedness is its small size which limits the information available (Fig. 3.11). Nevertheless, it is a useful indicator of relatedness as shown in the dendrogram in Fig. 3.12 which outlines the relationships amongst the Vibrionaceae and the Enterobacteriaceae based on 5S rRNA sequence comparisons.

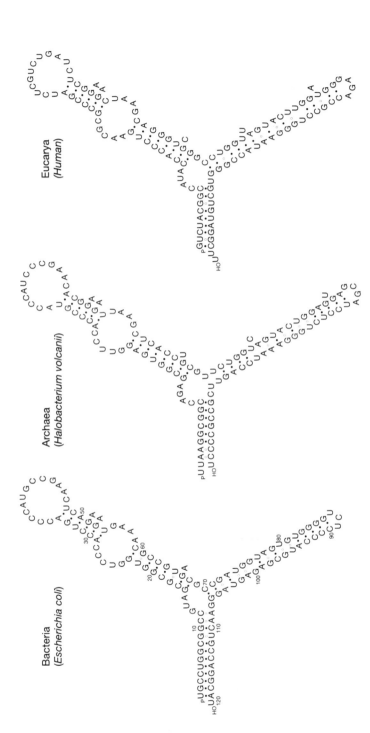

Fig. 3.11 A comparison of the 5S rRNA molecules from the three major evolutionary lines; Bacteria, Archaea and Eucarya. Note the general similarity in structure but numerous differences in sequence.

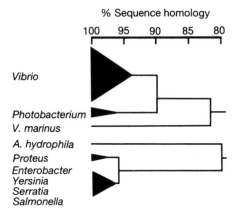

Fig. 3.12 Dendrogram indicating relationships among genera of the Enterobacteriaceae and Vibrionaceae based on 5S rRNA sequence data (from MacDonell and Colwell, 1984, with permission).

3.3.2   Sequence analysis of 16S rRNA

It can be argued that the small size of the 5S rRNA molecule detracts from its value in measuring relatedness between organisms, since it can undergo marked mutational change that would be obscured by the long stretches of conserved sequence present in the larger molecules. Largely for this reason, the 16S molecule, which is somewhat easier to handle than the 23S, has been used extensively for comparative sequencing studies (Woese, 1987). Prior to 1985, the 16S rRNA or its gene was too large to sequence in its entirety and so a cataloguing approach was adopted. This process involved digesting purified 16S rRNA with T1 ribonuclease into oligonucleotides. The 5′-termini of these molecules were labelled *in vitro* with ^{32}P, and they were separated by thin layer chromatography (TLC). This provided an oligonucleotide 'fingerprint' of the rRNA and each oligonucleotide was then sequenced to produce a 'catalogue' of sequences. Taxonomic structure was derived from catalogues of organisms, by comparing each catalogue with every other catalogue. Only oligonucleotides of six residues or more, common to two catalogues, were considered and an estimate of similarity was calculated using a Dice-type coefficient defined as:

$$S_{AB} = 2N_{AB} / (N_A + N_B)$$

where N_{AB} is the total number of residues in common among the oligonucleotides from the two organisms A and B; N_A is the total number of residues in oligonucleotides from organism A; and N_B is the total number of residues in the oligonucleotides from organism B. Thus S_{AB} values, like other coefficients of affinity, range from 1.0 for

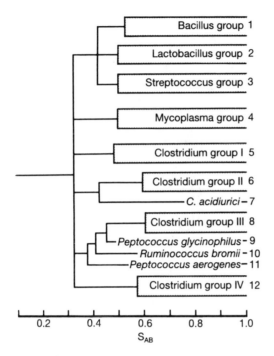

Fig. 3.13 Dendrogram of relationships among Gram-positive bacteria based on comparative cataloguing of 16S rRNA (adapted from Stackebrandt and Woese, 1981). *Bacillus* group: *Bacillus* species, *Peptococcus saccharolyticus, Planococcus citreus, Sporolactobacillus inulinus, Sporosarcina ureae, Staphylococcus* species, *Thermoactinomyces vulgaris. Lactobacillus* group: *Kurtha zopfii, Lactobacillus* species, *Leuconostoc mesenteroides, Pediococcus pentosaceus. Streptococcus* group: *Streptococcus* species. *Mycoplasma* group: *Acholeplasma laidlawii, Clostridium innocuum, C. ramosum, Mycoplasma capricolum, M. galli, Spiroplasma citri. Clostridium* group II: *C. latuseburense, Eubacterium tenue, Peptostreptococcus anaerobius. Clostridium* group III: *C. sphenoides, C. aminovalericum. Clostridium* group IV: *C. barkeri, Acetobacterium woodii, Eubacterium limosum.*

identical molecules to about 0.03 for totally unrelated molecules. The S_{AB} values were arranged as a similarity matrix and analysed by standard clustering procedures (see Chapter 2), usually average linkage analysis, to produce a dendrogram (Fig. 3.13). We emphasize that this dendrogram shows the present day relationships of organisms based on the 16S rRNA sequence and is a phenetic classification, just as a numerical taxonomy analysis of the phenotype is phenetic. Phylogenetic inferences could be extracted from this dendrogram if so desired (see Chapter 5) but the classification is phenetic.

The 16S rRNA can now be sequenced using various approaches. Historically, cataloguing was superseded by reverse transcriptase sequencing in which the 16S rRNA from cells was used as the template for dideoxy sequencing reactions. Rather than using DNA polymerase, reverse transcriptase (RNA dependent, DNA polymerase) was used to generate the dideoxy-terminated, deoxyribooligonucleotides. These were then separated by gel electrophoresis as in standard DNA sequencing procedures (Lane, 1991). As sequence data accumulated and were analysed, Lane *et al.* (1985) described a series of strategically positioned universally conserved primers that could be used to sequence virtually any small subunit rRNA. The annealing positions of these primers are shown on the diagram of *E. coli* RNA in Fig. 3.14.

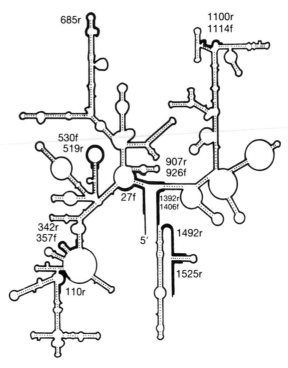

Fig. 3.14 Structure of the *E. coli* 16S rRNA molecule showing the positions of the 'universal' primers used for sequencing 16S rRNA molecules or their DNA equivalents (from Lane, 1991, with permission).

More recently, the polymerase chain reaction (PCR) has been used to generate amplified rRNA genes (Fig. 3.15). Primers to the extremities of the gene are used to amplify the DNA. The amplified DNA can either be sequenced directly or cloned into a plasmid or phage vector prior to sequencing. This is now the method of choice for most rRNA gene sequencing since it provides a straightforward route to unambiguous DNA sequencing.

Having generated the sequences, they are aligned so that nucleotide sites in correspondence are being compared. Simple comparisons of

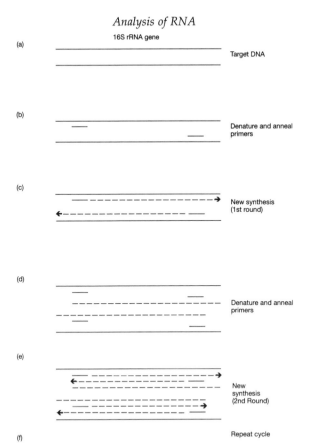

Fig. 3.15 The polymerase chain reaction for amplification of DNA sequences. (a) Chromosomal DNA containing 16S rRNA genes; (b) The DNA is denatured by high temperature and primers complementary to the ends of the gene are annealed in place; (c) New synthesis is generated from these primers (broken lines); (d) The duplicated molecules are denatured and the primers again annealed in place; (e) New synthesis is followed by repeated cycles.

sequence positions will provide a phenetic estimate of relatedness but most phylogenetic comparisons attempt to account for mutation rates including estimates of back mutations etc. (see Chapter 5). As we mentioned above, UPGMA clustering will provide a phenetic classification but again, phylogenetic analyses must be used to infer evolutionary pathways.

3.3.3 Sequence analysis of 23S rRNA

The higher information content of the large subunit rRNA makes it an attractive marker for phylogenetic reconstruction, for confirming trees based on the small subunit molecule, and for resolving problems and

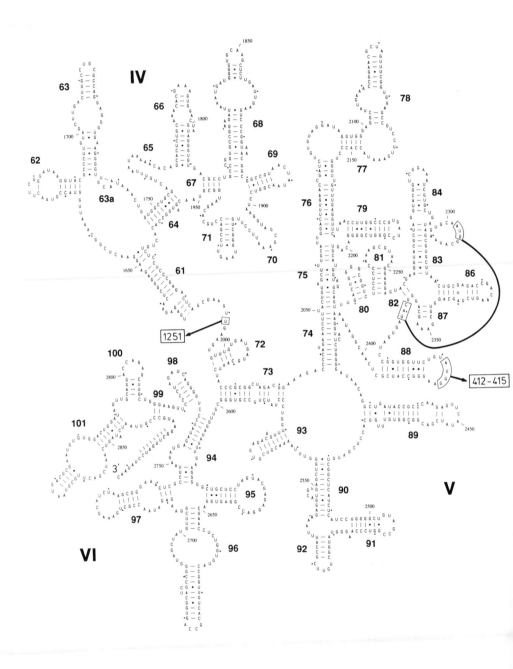

Fig. 3.16 Putative secondary structure model of the 23S rRNA of *Pseudomonas cepacia*. Roman numbers indicate domains; bold-faced arabic numbers indicate helices. Watson–Crick pairs are connected by thin lines. G·U and A·G pairings by filled and open circles, respectively. Putative tertiary structure interactions are

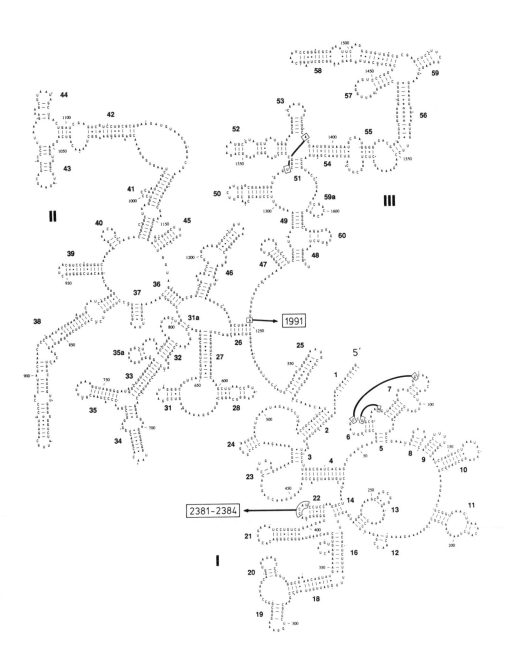

either joined by thick lines or their contacting bases are indicated. The 'sticky ends' at 5′ and 3′ termini illustrate how the ends should be joined in a eubacteria-specific proven helix (from Hopfl *et al.*, 1989, with permission).

discrepancies. The secondary structure of the 23S rRNA exhibits six major domains compared with three in the 16S molecule. Moreover, there are about 100 helices in the 23S molecule but only 50 in the 16S rRNA (Fig. 3.16). Both molecules have mosaics of conserved and gradually less conserved regions. This is apparent in the generalized secondary structure model of Proteobacter 23S rRNA shown in Fig. 3.16 in which bases which are conserved universally among bacteria are shown in capital letters and less conserved bases by dots. Indeed, in comparison with 16S rRNA there is more variety in the 23S rRNA sequence.

The 23S rRNA can be sequenced directly using the reverse transcriptase method but its larger size, greater level of base modification and lack of conserved primer sites complicates the procedure. PCR amplification followed by direct sequencing of the product or cloning into a vector are preferable for this more complex molecule.

3.3.4   DNA:rRNA hybridization

Ribosomal RNA is transcribed from about 10 cistrons in the chromosome. In DNA:rRNA hybridizations, the sequence homology between labelled 16S or 23S rRNA from a reference strain, and the rRNA cistrons within the chromosomal DNA from a second organism, are determined. This estimate of homology can be further analysed by estimating the extent of mismatched bases from the depression in T_m of the hybrid (see section 3.2.4).

The usual approach is to immobilize denatured chromosomal DNAs (about 50 μg) from a range of bacteria on nitrocellulose filters. The filters are incubated with radioactively labelled rRNA (10 μg) at optimal hybridization temperature for 16 h. The filters are then washed, treated with RNAase, and the total amount of labelled RNA bound (as μg RNA per 100 μg DNA), is the '% rRNA binding'. The thermal stability of the hybrid is estimated by measuring the release of label as the temperature of the filter is increased, and the $T_{m(e)}$ is the temperature at which 50% of the hybrid is eluted.

Taxonomic relatedness is revealed in two ways. Similarity maps, in which the ordinates are rRNA binding and $T_{m(e)}$, reveal clusters of organisms and show overlap between groups but do not provide a hierarchical classification. Where many reciprocal $T_{m(e)}$ determinations have been performed, the data can be treated as in a similarity matrix and clustered using the usual algorithms to provide a dendrogram that indicates the relationships in a hierarchical fashion (Fig. 3.17).

DNA:rRNA hybridization has proved to be an important tool for the elucidation of bacterial relationships at the intergeneric and suprageneric levels and there is excellent agreement between results obtained from DNA:rRNA hybridization and 16S rRNA cataloguing

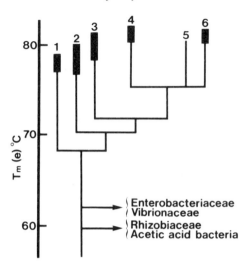

Fig. 3.17 Dendrogram of relationships among some Gram-negative bacteria based on thermal stabilities of rRNA:DNA hybrids. The shaded areas represent the range of $T_{m(e)}$ values of the reference taxon. 1, *Janthinobacterum lividum*; 2, *Chromobacterium violaceum*; 3, *Pseudomonas solanacearum*; 5, *Alcaligenes denitrificans*; 6, *Alcaligenes faecalis* (adapted from De Ley *et al.*, 1978).

and sequencing studies (Stackebrandt, 1988). Many strains can be examined rapidly because hybridization reactions are less time-consuming than sequencing. On the negative side, as with DNA reassociation assays, DNA:rRNA hybridization studies do not generate a full matrix of homology values and the results can be affected by unfortunate or inappropriate choice of reference strain. Secondly, it is difficult to make comparisons across very wide taxonomic boundaries because of lack of sequence homology. For these reasons it seems likely that the hybridization approach will be secondary to sequencing in most applications.

3.3.4 Value and applications of rRNA studies

Ribosomal RNA analysis is an invaluable procedure for extending classifications, established by DNA reassociation and phenotypic studies, to more distant organisms, and is being used to provide a comprehensive view of the relationships among all prokaryotes and of the eubacteria to other kingdoms. Of the two methods, sequencing the molecules has the advantage of providing information for individual organisms that can be processed using estimates of similarity and clustering algorithms and steadily accumulated into a universal database. Hybridization studies, on the other hand, suffer the disadvantage noted above that they do not generate data for an individual strain, but compare sequences between reference and test strains. Moreover, hybridization is not as accurate as sequencing, particularly amongst distantly related organisms. Nevertheless, it is rapid, relatively straightforward and has provided valuable insight into the relationships amongst several Gram-negative and Gram-positive taxa (Gillis and De Ley, 1980; Grimont, 1988).

The application of rRNA analysis is having a dramatic effect on our more traditional views of the relationships among the prokaryotes and between prokaryotes and eukaryotes. Phenograms based on rRNA sequences and rRNA:DNA hybrids are providing:

1. A comprehensive overview of the classification of prokaryotes and indicating relationships that, on phenotypic grounds, had not been apparent.
2. Unifying concepts of the genus and higher-ranked taxa.
3. An appreciation of possible evolutionary pathways among all living (and fossilized) organisms.

We shall deal with these three points in order but shall reserve discussion of evolutionary pathways for Chapter 5.

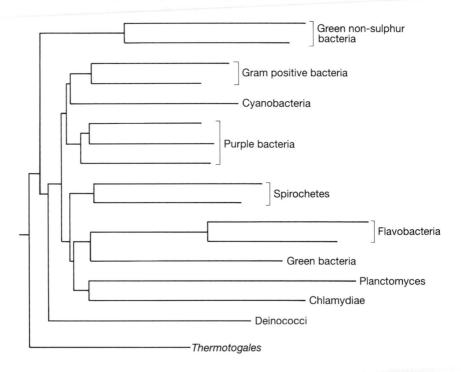

Fig. 3.18 Phylogenetic tree for the Bacteria based on 16S rRNA sequences. The alignment used comprised one to three representative sequences from each of the 11 recognized bacterial phyla (from Woese 1992, with permission).

Table 3.3   Eubacterial phyla and their major subdivisions

Proteobacteria (purple bacteria)

        α subdivision
        α-1   *Rhodospirillum*
        α-2   *Rhizobium*
        α-3   *Rhodobacter*
        α-4   *Erythrobacter*

        β subdivision
        β-1   *Pseudomonas testoteroni*
        β-2   *Chromobacterium violaceum*
             Pseudomonas cepacia
        β-3   *Neisseria gonorrhoea*
             Spirillum volutans

        γ subdivision
        γ-1   *Chromatium*
        γ-2   *Legionella*
        γ-3   Fluorescent pseudomonads
             Enterics, *Aeromonas, Vibrio*

        δ subdivision
        sulphur and sulphate reducers

Gram-positive eubacteria

        A   High G and C species
             Actinomyces, Micrococcus, Streptomyces
        B   Low G and C species
             Clostridium, Bacillus, Lactobacillus
             Mycoplasmas
        C   Photosynthetic species
             Heliobacterium
        D   Species with Gram-negative walls
             Megasphaera

Cyanobacteria

        Aphanocapsa, Oscillatoria, Nostoc

Spirochaetes and relatives

        A   Spirochaetes
             Treponema, Borrelia
        B   *Leptospiras*

Green sulphur bacteria

        Chlorobium, Chloroherpeton

Bacteroides, Flavobacteria and relatives

        A   Bacteroides
             Bacteroides, Fusobacterium
        B   Flavobacterium group
             Cytophaga, Flavobacterium

continued

Table 3.3 *continued*

Planctomyces and relatives

    A Planctomyces group
       Planctomyces, Pasteuria
    B Thermophiles
       Isocystis

Chlamydia

    Chlamydia

Radioresistant micrococci and relatives

    A Deinococcus group
       Deinococcus radiodurans
    B Thermophiles
       Thermus aquaticus

Green non-sulphur bacteria and relatives

    A Chlorflexus group
       Chlorflexus, Herpetosiphon
    B Thermomicrobium group
       Thermomicrobium

Thermotoga

This table includes representative examples of the 11 groups, for a more complete listing see Woese (1987) and the appendices.

(a) Overview of bacterial classification

The early days of rRNA cataloguing and the later studies involving complete 16S rRNA sequences (which incidentally did not change the major conclusions of the cataloguing work) have been used to divide the eubacteria into 11 major divisions or phyla (Table 3.3, Fig. 3.18). These phyla may contain some unexpected members with morphologically and physiologically diverse genera included within a subdivision. For example the cell-wall deficient mycoplasmas are included with the clostridia in one of the Gram-positive phyla, and the fluorescent pseudomonads with enteric bacteria and *Legionella* in the δ subdivision of the proteobacteria. But most of the taxa are defined by unique oligonucleotide sequences or 'signatures' and each is coherent on the basis of rRNA sequence. It may be that, given this framework, unifying phenotypic features will be discovered so that the taxa can be defined phenetically as well as phylogenetically. Alternatively, phenotypic markers may be recognized as poor indicators of phylogeny and abandoned to the oligonucleotide signatures; a prospect which could cause some argument and confusion (Sneath, 1989). Nevertheless, we provide here a summary of the composition and physiologies of the major phyla of the eubacteria as revealed by rRNA sequence comparisons.

Proteobacteria (purple bacteria). More than 200 genera of Gram-negative bacteria are included in the proteobacteria or purple bacteria but the arrangements of the classically defined genera and species are mixed up. At least four distinct subdivisions have been recognized (*Campylobacter* may comprise a fifth) with photosynthetic purple bacteria forming the nucleus of three of the subdivisions (hence the emphasis on these bacteria in the original name). Since photosynthesis is so complex it is unlikely to have arisen more than once, the ancestor of this group is therefore thought to be a purple photosynthetic bacterium. Photosynthetic capacity was subsequently lost several times in this phylum and replaced by other physiological processes, such as sulphate reduction and lithotrophy; features better adapted to new and expanding ecological niches. The α and β subdivisions contain non-sulphur purple bacteria while purple sulphur bacteria are located in the γ subdivision. The γ subdivision also contains the genera *Aeromonas*, *Pseudomonas* and *Vibrio* and the family Enterobacteriaceae. The δ subdivision is diverse and contains bacteria such as the strictly anaerobic sulphate reducers, the fruiting myxobacteria and the prokaryotic predator *Bdellovibrio*, a small spiral shaped Gram-negative bacterium which is a predator of other Gram-negative bacteria.

Gram-positive eubacteria. With the exception of *Deinococcus*, the Gram-positive bacteria form a relatively homogeneous line of descent (Fig. 3.19). A major partition divides the high G+C content actinomycetes and relatives from the other groups. This phylum contains some unexpected members such as the micrococci, which have previously been accommodated rather awkwardly with *Staphylococcus*. The assignment of these cocci to the actinomycete branch is supported by other chemotaxonomic markers, such as cell-wall peptidoglycan structure. The ancestor of the low G+C branch, which includes *Bacillus*, *Clostridium*, *Lactobacillus*, *Ruminococcus*, *Staphylococcus* and *Streptococcus* was probably an anaerobic endospore-forming bacterium. Genes for the small acid-soluble spore proteins, which are a major component of the endospore and responsible for resistance to ultraviolet irradiation, are so highly conserved and the spore is so complex that sporulation can only have evolved once and genera such as *Staphylococcus* and *Streptococcus* have lost the capacity to differentiate into endospores. The filamentous endospore-forming bacterium *Thermoactinomyces* is an interesting genus that has traditionally been classified with the actinomycetes by virtue of its substrate and aerial mycelia, but the presence of endospores, relatively low G+C content of the chromosomal

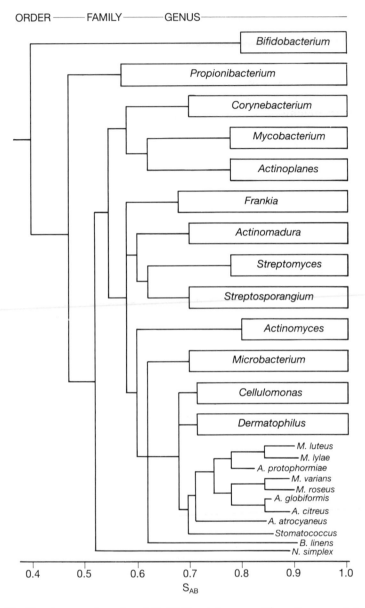

Fig. 3.19 Dendrogram based on 16S rRNA catalogues of the actinomycete branch of the Gram-positive bacteria and their relatives (from Fox and Stackebrandt, 1987, with permission).

DNA and cell wall structure are all consistent with its inclusion in the family Bacillaceae; as confirmed by RNA sequence comparisons. The mycoplasmas are an interesting branch of the phylum which have lost the ability to synthesize a cell wall and are bounded only by a membrane.

Previously thought to be derived from Gram-negative bacteria, it is now clear that the mycoplasmas have arisen from the clostridia.

Cyanobacteria. The classical blue-green algae are defined by the common possession of chlorophyll *a*. The branching and heterocystous forms of cyanobacteria appear to form discrete branches but the unicellular cyanobacteria are more diverse. The sequences of cyanobacterial and chloroplast rRNA genes have much in common indicating that they shared a common ancestor. The green plant chloroplast was probably not derived from a cyanobacterium itself, rather a common photosynthetic relative gave rise to both the chloroplast and the cyanobacterial lines.

Spirochaetes and relatives. The spirochaetes are one of the few phyla that have been recognized on both traditional criteria and by rRNA sequence comparisons. Their common spiral shape and axially coiled filaments set them apart from other bacteria. Two clearly separated subdivisions are apparent in the phylum; one composed of *Borrelia*, *Spirochaeta* and *Treponema* and the other *Leptospira*.

Green sulphur bacteria. The photosynthetic green sulphur genera; the non-motile *Chlorobium* and the gliding bacteria *Chloroherpeton* branch from the main eubacterial line at a very early stage. The unique pigments and light-harvesting apparatus (chlorosomes) of these organisms are in agreement with this ancient lineage.

Bacteroides, flavobacteria and relatives. Several genera form this major phylogenetic line of Gram-negative bacteria. One subline comprises the anaerobic *Bacteroides* and *Fusobacterium*, while the second contains aerobic genera such as *Cytophaga*, *Flexibacter* and *Saprospira*, all rod-shaped or filamentous organisms, primarily respiratory in physiology and which move by gliding motility.

Planctomyces and relatives. This phylum is defined by species variously placed in the genera *Pasteuria*, *Pirella* and *Planctomyces* and the hot spring organism *Isocystis pallida*. All these organisms lack peptidoglycan and possess a proteinaceous cell wall. In terms of oligonucleotide sequence, this is a highly unusual group and it shares very low sequence homology with other eubacterial phyla. The organisms contain few of the conserved (ancient) sequences characteristic of the eubacteria and the current opinion is that they have evolved far more rapidly than other lineages.

Chlamydiae. The obligate, intracellular parasites of the genus *Chlamydia* constitute a separate phylum. Like the planctomycete group, these bacteria lack peptidoglycan and they have oligonucleotide signatures in common with this group. Too few species have been studied to describe this phylum adequately.

Radiation-resistant micrococci and relatives.   This phylum consists of two well defined, but rather different genera. *Deinococcus* includes the highly radiation-resistant coccus, *D. radiodurans* and its relatives. The ubiquitous, hot-spring organism *Thermus aquaticus* forms the second branch within the phylum. Interestingly, both *Deinococcus* and *Thermus* have an atypical cell wall in which diaminopimelic acid in the peptidoglycan is replaced by ornithine.

Green, non-sulphur bacteria and relatives.   This phylum is one for which little phenotypic justification exists at present. Initial studies of the green, non-sulphur bacterium *Chlorflexus* suggested a close relationship to the green sulphur bacteria, in particular chromosome structure and light-harvesting pigment type, but from sequence comparisons these genera are very different and the structures of their photoreaction centres differ. The *Chlorflexus* group includes the thermophilic prototroph *C. auriantiacum*, two non-photosynthetic genera; *Herpetosiphon* and *Thermomicrobium*, which are mesophilic, gliding bacteria and the thermophile *Thermomicrobium roseum*. The unusual long chain diols found in *Thermomicrobium*, which are the functional equivalent of normal glycerol lipids, have also been identified in *Chlorflexus* indicating a chemotaxonomic bond between the genera.

Thermotoga.   The final group of eubacteria consists of a single genus *Thermotoga* which contains one species *T. marina*. This highly thermophilic, strictly anaerobic, fermentative bacterium is capable of growth at temperatures of 55–90°C with an optimum of about 80°C. It is the most thermophilic of all eubacteria. *Thermotoga* contains unique, long chain fatty acids and its cell wall contains peptidoglycan. Although clearly a eubacterium, *T. marina* is positioned a long way from other eubacteria in the tree (Fig. 3.17) and it may represent a nucleus of a large domain of thermophilic eubacteria with unusual properties and biochemistry.

The 11 groups described above represent most of the eubacteria examined by rRNA cataloguing or sequencing to date (Stackebrandt, 1992). Given that we have probably cultured a minority of extant bacteria (perhaps only 1–5%), it would seem likely that other phyla will be defined as we explore the biodiversity on this planet more thoroughly.

(b) Unifying concepts of the genus and higher ranked taxa

Prior to the introduction of molecular systematics many genera were poorly defined. The vast differences in base composition in many genera

(Table 3.1) attest to the variation in genetic structure encompassed by genera such as *Clostridium* and *Lactobacillus*. Can phylogenetic studies based on sequence comparisons help to define genera and higher taxa more naturally? It is important to realize that no single level of relatedness can be used to define a genus in the same way as a species description can be formulated on the basis of DNA sequence homology. It is not possible to draw a line across a cladogram and nominate all groups above the line as genera. If such an attempt was made, at an S_{AB} level of about 0.8 (the point at which proteobacterial Gram-negative genera are separated) many Gram-positive species, which show lower branching levels than the Gram-negative species, would become genera. If, on the other hand, a value of S_{AB} of about 0.7 was used to divide the Gram-positive organisms into acknowledged genera, many Gram-negative genera would be united into one genus.

This problem has been addressed by Stackebrandt (1988) who explained why such an inflexible approach to the use of sequence similarities for the circumscription of genera was inappropriate. Phenotypically defined genera are of varying ages and in general can be assigned to one of three age groups. The first age group contains ancient genera. These have low sequence similarities (S_{AB} values less than 0.4) and the ancestral forms of these genera evolved during the anaerobic phase of evolution. Indeed, their descendants still live in niches which may reflect the ancient conditions. Representative genera include *Bacteroides*, *Clostridium*, *Eubacterium* and the spirochaetes. Some, such as the *Bacteroides* are coherent taxa, other, such as the clostridia are heterogeneous and comprise several genus-ranked taxa.

The second age group comprises genera whose members evolved during the initiation of an aerobic atmosphere on the planet. The extant forms of these bacteria include *Actinomyces*, *Bacillus*, *Lactobacillus* and other Gram-positive genera. The S_{AB} values of these genera are about 0.45. Again, many of these genera are taxonomically incoherent and in need of redefinition, indicating that the phenotypic descriptions are insufficient.

The third age group is larger and contains genera that evolved during the aerobic phase of the biosphere. The defining S_{AB} values for these genera, which include the four subdivisions of the proteobacteria, most of the high G+C content Gram-positive bacteria and some other phyla, are about 0.7. These genera are usually well defined phenotypically.

Most, but not all, of the eubacteria fit into these three age categories and it will be apparent that the number of genera adequately defined by both rRNA sequence relationships and phenotypic characters is small. Of the ways forward, some favour a classification which is entirely cladistic, as derived from sequence studies, and completely independent of phenotypic characters. Such a special purpose

Nucleic acid analyses

classification (a Hennigean reference system; see Chapter 1) may not serve microbiology very usefully, particularly where phylogenetically coherent taxa are phenetically heterogeneous. The adherents of such schemes stress that when phylogenetically circumscribed taxa are disclosed, definitive phenotypic characters are subsequently detected. For example, the endospores of *Thermoactinomyces* and *Bacillus* or the lack of peptidoglycan in the planctomycetes. This is encouraging, but there seems little to be gained from persevering with phylogenetically defined taxa which steadfastly remain phenetically heterogeneous. Conversely, if extreme convergence due to strong selection pressures leads to two phylogenetically distinct lineages becoming phenetically identical, there is little practical value in separating them and the phylogenetic classification is of little value outside the restricted field of genealogy. Classification also has the responsibility of providing identification schemes and information databases for microbiologists in general and these more prosaic demands must not be forgotten in the dream of reconstructing the evolution of the microbial world.

A more flexible approach has been adopted by Stackebrandt (1988), which largely concurs with the findings of an ad hoc committee convened to report on the reconciliation of these two approaches to bacterial systematics (Wayne *et al.*, 1987). It is effectively the position of the 'traditional evolutionist' (see Chapter 1) in which phylogenetic relationships are used but consistency with phenetic relationships is emphasized. Thus Fox and Stackebrandt (1987) suggest sequence relationships should be used to outline the skeleton of the classification and ranks are defined phylogenetically, but the ranges are adjusted to be consistent with phenotypic characters. On an optimistic note, Stackebrandt (1988) suggests that as new characters are found, the phenotypic system will be subsumed by the phylogenetic and a single classification will finally emerge. Only time will reveal if the single, multipurpose, stable classification can finally be achieved.

REFERENCES

Bradley, S.G. and Mordarski, M. (1976). Association of polydeoxyribonucleotides of deoxyribonucleic acids from nocardioform bacteria, in *Biology of the Nocardiae*, (M. Goodfellow, G.H. Brownell and J.A. Serrano, Eds.), pp. 310–336. Academic Press, London.

Brenner, D.J. (1970). Deoxyribonucleic acid divergence in Enterobacteriaceae, *Developments in Industrial Microbiology*, **11**, 139–153.

Britten, R.J. and Kohne, D.E. (1966). Nucleotide sequence repetition in DNA, *Carnegie Institute Yearbook*, **65**, 78–106.

Brown, T.A. (1992). *Genetics; A Molecular Approach*, 2nd edition, Chapman & Hall, London.

De Ley, J., Segers, P. and Gillis, M. (1978). Intra- and intergeneric similarities of *Chromobacterium* and *Janthinobacterium* ribosomal ribonucleic acid cistrons, *International Journal of Systematic Bacteriology*, **28**, 154–168.

De Muro, M.A., Mitchell, W.J. and Priest, F.G. (1991). Differentiation of mosquito pathogenic strains of *Bacillus sphaericus* from non-toxic varieties by ribosomal RNA gene restriction patterns, *Journal of General Microbiology*, **138**, 1159–1166.

Fox, G.E. and Stackebrandt, E. (1987). The application of 16S rRNA cataloguing and 5S rRNA sequencing in bacterial systematics, *Methods in Microbiology*, **19**, 405–458.

Gillis, M. and De Ley, J. (1980). Intra- and intergeneric similarities of ribosomal ribonucleic acid cistrons of *Acetobacter* and *Gluconobacter*, *International Journal of Systematic Bacteriology*, **30**, 7–27.

Grimont, P.A.D. (1988). Use of DNA reassociation in bacterial classification, *Canadian Journal of Microbiology*, **34**, 541–546.

Grimont, F. and Grimont, P.A.D. (1991). DNA fingerprinting, in *Nucleic Acid Techniques in Bacterial Systematics* (E. Stackebrandt and M. Goodfellow, Eds.), pp. 249–280, Wiley, Chichester.

Höpfl, P., Ludwig, W., Schleifer, K.H. and Larsen, N. (1989). The 23S ribosomal RNA higher-order structure of *Pseudomonas cepacia* and other prokaryotes, *European Journal of Biochemistry*, **185**, 355–364.

Johnson, J.L. (1991). DNA reassociation experiments, in *Nucleic Acid Techniques in Bacterial Systematics* (E. Stackebrandt and M. Goodfellow, Eds.), pp. 21–44, Wiley, Chichester.

Krych, V., Johnsen, J.L. and Yousten, A.A. (1980). Deoxyribonucleic acid homologies among strains of *Bacillus sphaericus*, *International Journal of Systematic Bacteriology*, **30**, 476–484.

Lane, D.J. (1991). 16S/23S rRNA sequencing, in *Nucleic Acid Techniques in Bacterial Systematics* (E. Stackebrandt and M. Goodfellow, Eds.), pp. 115–176, Wiley, Chichester.

Lane, D.J., Pace, B., Olsen, G.J. *et al.* (1985). Rapid determination of 16S ribosomal RNA sequences for phylogenetic analysis, *Proceedings of the National Academy of Sciences of the United States of America*, **82**, 6955–6959.

Ludwig, W. (1991). DNA sequencing in bacterial systematics, in *Nucleic Acid Techniques in Bacterial Systematics* (E. Stackebrandt and M. Goodfellow, Eds.), pp. 69–94, Wiley, Chichester.

MacDonell, M.T. and Colwell, R.R. (1984). The nucleotide sequence of 5S ribosomal RNA from *Vibrio marinus*, *Microbiological Sciences*, **1**, 229–231.

Nei, M. and Lei, W.-H. (1979). Mathematical model for studying genetic variation in terms of restriction endonucleases. *Proceedings of the National Academy of Sciences of the United States of America*, **76**, 5269–5273.

Owen, R.J. and Pitcher, D. (1985). Current methods for estimating DNA base composition and levels of DNA-DNA hybridization, in *Chemical Methods in Bacterial Systematics* (M. Goodfellow and D.E. Minnikin, Eds.), pp. 67–93, Academic Press, London.

Priest, F.G. (1992). Biological control of mosquitoes and other biting flies using *Bacillus sphaericus* and *Bacillus thuringiensis*: A review. *Journal of Applied Bacteriology*, **72**, 357–369.

Rodrigues, U.M., Aguirre, M., Facklam, R.R. and Collins, M.D. (1991). Specific and interspecific molecular typing of lactococci based on polymorphism of DNA encoding rRNA, *Journal of Applied Bacteriology*, **71**, 509–516.

Sharp, P.M., Nolan, N.C., Cholmain, N.N. and Devine, K.M. (1992). DNA sequence variability at the *rpl X* locus of *Bacillus subtilis*, *Journal of General Microbiology*, **138**, 39–45.

Sneath, P.H.A. (1989). Analysis and interpretation of sequence data for bacterial systematics: the view of a numerical taxonomist, *Systematic and Applied Microbiology*, **12**, 15–31.

Stackebrandt, E. (1988). Phylogenetic relationships vs. phenotypic diversity: how to achieve a phylogenetic classification system of the eubacteria, *Canadian Journal of Microbiology*, **34**, 552–556.

Stackebrandt, E. (1992). Unifying phylogeny and phenotypic diversity, in *The Prokaryotes*, 2nd edition (A. Balows, H.G. Trüper, M. Dworkin, W. Harder and K.M. Schleifer, Eds.), pp. 19–46, Springer-Verlag, New York.

Stackebrandt, E. and Woese, C.R. (1981). The evolution of prokaryotes, *Symposium of the Society for General Microbiology*, **32**, 1–32.

Stahl, D.A. and Amann, R. (1991). Development and application of nucleic acid probes, in *Nucleic Acid Techniques in Bacterial Systematics* (E. Stackebrandt and M. Goodfellow, Eds.), pp. 205–248, Wiley, Chichester.

Staley, T.E. and Colwell, R.R. (1973). Application of molecular genetics and numerical taxonomy to the classification of bacteria, *Annual Review of Ecology and Systematics*, **4**, 273–300.

Wayne, L.G., Brenner, D.J., Colwell, R.R. *et al.* (1987). Report of the ad hoc committee on reconciliation of approaches to bacterial systematics, *International Journal of Systematic Bacteriology*, **37**, 463–464.

Woese, C.R. (1987). Bacterial evolution, *Microbiological Reviews*, **51**, 221–271.

Woese, C.R. (1992). Prokaryote systematics: the evolution of a science, in *The Prokaryotes*, 2nd edition (A. Balows, H.G. Trüper, M. Dworkin, W. Harder and K.M. Schleifer, Eds.), pp. 1–8, Springer-Verlag, New York.

CHAPTER 4

Chemosystematics and molecular biology II: Proteins, lipids, carbohydrates and whole cells

The current preoccupation with the molecular biology of the nucleic acids has tended to detract from the immense contribution that the chemistry of other cell constituents can make to bacterial classification and identification. Proteins, lipids, cell wall constituents and even whole cells all provide valuable chemical data for the distinction of bacterial species and genera, and can be used for both classification and identification often by mechanizing the analysis and combining it with computerized data handling for rapid automated identification. In some instances there is also excellent agreement between chemical structure and phylogeny, as for example in the case of cell wall composition of the Gram-positive bacteria.

An outline of the common chemical analyses used in chemosystematics and the levels at which they contribute to bacterial classification and identification is given at the beginning of Chapter 3 in Table 3.1.

4.1 ANALYSIS OF PROTEINS

Measurement of relationships between organisms using proteins focuses on either comparisons of individual molecules (using amino acid sequences or serological cross-reactions) or gross evaluation of total cellular protein. A refinement of the latter approach is to compare the electrophoretic mobilities of sets of enzymes (multilocus enzyme electrophoresis). These approaches and some example applications will be described below, with the exception of protein sequence analysis which is covered in Chapter 5.

4.1.1 Comparative serology of proteins

The development of quantitative serological techniques has enabled the rapid detection of structural similarities in isologous enzymes from different bacteria. A common approach is to use the chosen purified enzyme from several reference strains to raise antisera. These antisera are used to detect the isologous enzyme in cell extracts from test organisms. Two-dimensional immunodiffusion (Ouchterlony technique) is used to establish gross similarities between enzymes, a line of identity indicating closely related molecules, and crossed precipitin lines revealing more distantly related enzymes. The degree of relatedness can be estimated using microcomplement fixation tests in which the amount of complement fixed by homologous and heterologous mixtures can be used to quantify the reaction. The figures obtained can be placed in a similarity matrix and analysed using the usual clustering techniques.

It is generally assumed that sequence variation in proteins will be reflected in their secondary structures and, therefore, antigenicity. Quantitative serology is therefore a rapid approach to the analysis of protein structure in much the same way as rRNA:DNA hybridization is a rapid approach to rRNA sequencing. However, it has not been a particularly popular technique. Most studies have centred on the lactic acid bacteria, *Lactobacillus*, *Pediococcus*, *Leuconostoc* and *Streptococcus*, and the enzymes fructose (bisphosphate) aldolase, and glucose-6-phosphate-, glyceraldehyde-3-phosphate- and lactate dehydrogenases. The results confirm the close relationship of these genera, and are largely in accord with classifications derived from other sources. It would seem that a particularly useful aspect of comparative serology is to provide the preliminary ground work, and to indicate taxa and proteins that can be analysed in detail using protein sequencing. Nevertheless, like rRNA:DNA reassociation, it is also a valuable approach to detecting molecular relationships in its own right.

4.1.2 Comparative electrophoresis of cellular proteins

The bacterial genome is largely devoted to the production of some 2000 proteins, which function either enzymically or structurally. When a bacterium is grown under carefully standardized conditions, this protein complement is essentially invariant. Electrophoresis of the total cellular proteins in polyacrylamide gels (PAGE) provides a partial separation in which individual bands mostly represent several proteins. However, this complex pattern is reproducible and represents a 'fingerprint' of the strain that can be used for comparative purposes (Fig. 4.1).

The original electrophoretic systems involved polyacrylamide rod gels used in non-denaturing conditions, but more recently sodium

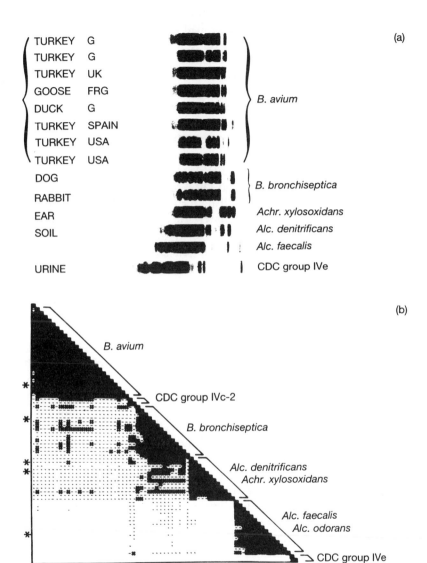

Fig. 4.1 (a) Normalized electrophoregrams of soluble proteins from eight avian *Bordetella*-like strains (*B. avium*) and representative strains of various more or less allied bacteria. The origin of the strains is shown at the left side of the figure. The avian strains were isolated in the former Federal Republic of Germany (FRG), United Kingdom (UK), Spain, and the United States of America (USA). (b) Sorted similarity matrix from the numerical analysis of protein patterns of 24 avian *Bordetella*-like strains (*B. avium*) and various related strains: ■, 95–100% similarity; ◨, 90–94%; #, 85–89%; —, 80–84%; ●, 70–79%; □, 0–69% (from Kersters, 1985, with permission).

dodecyl sulphate (SDS) PAGE has found greater application. The bacteria are broken down, usually by physical means, and the lysate applied directly to the gel. One or two reference proteins are included and, after electrophoresis, the gel is stained with, for example, Coomassie brilliant blue. A densitometer trace of the stained gel provides the quantitative data for the bacterium.

Analysis of the densitometer traces uses traditional numerical methods. The peak heights (absorbances) are normalized using the internal reference proteins after removal of background 'noise', and used as the characters for the organism. Similarities are calculated between each organism using a suitable similarity coefficient; the Pearson product–moment correlation and Dice coefficients have been popular. The resultant matrix is clustered using the average linkage algorithm to provide a sorted similarity matrix or dendrogram. Computer programs are readily available to perform these calculations (Jackman, 1985) and the latest versions allow gels to be scanned by document scanners rather than densitometers (e.g. Gel Manager for Windows from BioSystematica, Prague).

Comparative electrophoresis of proteins should be equivalent to DNA:DNA reassociation, since bacteria are being compared on the translational products of most of the cell's chromosome. It is to be expected, therefore, that it will be most useful at the species level, and of little value when comparing distantly related bacteria. This is the experience to date. The two methods give highly congruent results. Protein electrophoresis has several advantages over DNA:DNA pairing. Clusters are formed from complete similarity matrices, since a set of data is produced for each strain rather than comparisons between reference and test strains. The technique is more rapid than DNA reassociation, particularly since 20 samples can be electrophoresed in a single slab gel, there are no lengthy preparation procedures and the data can be entered directly from the densitometer into a computer. Thirdly, it is more amenable to identification than DNA:DNA reassociation. Patterns for reference clusters can be held on disc, and patterns for unknown organisms can be compared with the data bank in order to provide an identification, in much the same way as probabilistic methods are used for identification based on phenotypic characters (see Chapter 7). However there are some drawbacks. Although reproducibility within the laboratory does not seem to be a serious problem, inter-laboratory comparisons may not always be entirely reliable. The adoption of standard procedures may improve reproducibility. Although any group of bacteria can be studied, standardization of growth conditions may be difficult or impossible when comparing different physiological types.

Despite the advantages of the system, comparative electrophoresis of proteins has not been widely adopted. It has been used in conjunction with traditional numerical taxonomy and DNA:DNA reassociation to delineate species of *Zymomonas* and for the classification and identification of *Alcaligenes, Achromobacter, Bordetella, Corynebacterium* and *Pseudomonas* species among others. In virtually all cases, there was excellent agreement between the three approaches (for review see Kersters, 1985).

4.1.3 Multilocus enzyme electrophoresis (MLEE)

Rather than an analysis of the total proteins in a cell, the electrophoretic properties of several enzymes can be compared. Such studies have focused on enzymes that are common to a group and can be readily detected in native (non-denaturing) electrophoresis gels through a specific reaction that results in colour formation (esterases and dehydrogenases have been popular; Selander *et al.*, 1986). Thus, strains can be classified into groups on the basis of the presence or absence of particular enzymes and by comparison of their electrophoretic mobilities. This approach is borrowed directly from population genetic studies of higher organisms and has been invaluable in understanding the population genetics of bacteria (Selander and Musser, 1990).

The rationale of the method lies in a comparison of minor changes in gene structure between organisms based on variation in enzyme structure. Obviously, it would be preferable to compare the gene sequences directly, but for the large numbers of organisms involved and the preference for examining numerous loci (routinely 15–25) DNA sequencing is still untenable. Instead, because the net electrostatic charge and, hence, migration of a protein during electrophoresis are determined by its amino acid sequence, mobility variants, sometimes called electromorphs or allozymes (but preferably not isozymes since this term includes the same type of enzyme coded by separate loci) of an enzyme can be directly equated with alleles at the corresponding structural gene locus. We are in effect comparing gene sequences as the expression of the gene, and in the form of allozymes have very fine tools with which to do this; much finer than DNA sequence homology determinations by reassociation assays. Indeed, for pairs of isolates of *E. coli, Legionella* and *Gluconobacter*, it has been demonstrated that estimates of genetic distance based on MLEE correlate strongly with those derived from reassociation experiments. The comparisons are made within the species. For example one may wish to determine the taxonomic and genetic structure of *B. subtilis* or *Haemophilus influenzae* but little would be gained from comparing *B. subtilis* with *B. cereus* or *H. influenzae* with *H. pleuropneumoniae*; the organisms of these two species would have diverged too greatly to obtain useful results. MLEE is used to gain an insight into the genetic structure of the species in a population genetics context, not to compare strains of different species.

In brief, the bacteria to be examined are cultured on a standard medium and a cell extract prepared. This can generally be frozen or examined directly by native gel electrophoresis. Some 20 cell extracts can be examined on a single slab gel which is stained after electrophoresis for a particular enzyme. Starch gels are still very popular because of the ease with which they can be handled, and agarose gels can also be used, although the lower resolution of this matrix often gives diffuse bands. Polyacrylamide gels usually give optimum results. Staining for enzyme activity uses colorimetric reactions, for example, dehydrogenases are detected by coupling the enzyme reaction to dimethylthiazol tetrazolium and phenazine methosulphate to produce a coloured reaction and esterases can be detected by using napthyl acetate and fast blue salt RR. Methods for the detection of large numbers of enzymes have been collated by Selander *et al.* (1986) and May (1992).

Comparisons of the enzyme mobilities are made visually against a standard strain included on the gel. Relative mobilities are not generally used, because of inaccuracies in their estimation, instead an electromorph is determined as a mobility variant of an enzyme and numbered in order of decreasing anodal migration. Other strains are examined for this particular enzyme and if the protein has exactly the same electrophoretic mobility it is considered to be the same electromorph. Different mobility variants are different electromorphs. Each isolate is therefore characterized by its combination of electromorphs. The data can be analysed by simply treating each electromorph as a unit character and from the presence/absence data constructing a similarity matrix using the Jaccard or similar coefficient which ignores negative matches (see Table 2.6). It may be beneficial to use special weighted coefficients which take into account the genetic diversity at the locus and give greater weight to differences at less variable loci than to those at highly variable loci (Selander *et al.*, 1986). When distance coefficients are used the data are considered as estimates of genetic distance. Clustering from the similarity or distance matrix is best achieved by the UPGMA algorithm to provide a dendrogram or multivariate statistics can be used to produce scatter diagrams.

Comprehensive MLEE studies of largely Gram-negative bacteria have led to three generalized conclusions (Selander and Musser, 1990).

1. Most bacterial species are clonal in structure. That is, the species comprises an assemblage of identical or near identical types which are called clones. Members of a clone have identical multilocus enzyme genotypes, are stable over long periods and can be recovered worldwide. Since it is highly unlikely that two isolates from different genetic stock would converge to identical multilocus genotype, these two isolates must have originated from the same parent. Generally, members of a clone can be isolated from all around the world,

indicating that the clone has become disseminated and is stable. Many species such as *E. coli, Haemophilus* spp., *Bordetella* spp. and legionellae have this strong clonal structure which is indicative of very little genetic recombination between clones. If gene transfer and recombination were common, the clonal structure should be eroded as alleles from other strains are accumulated in the

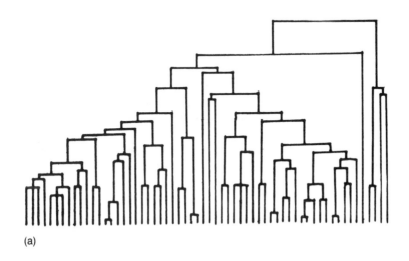

(a)

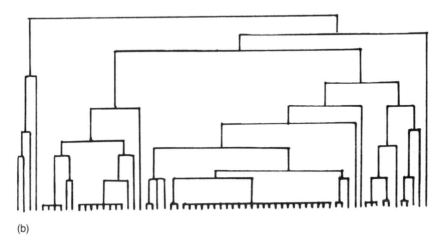

(b)

Fig. 4.2 Dendrograms showing hypothetical results from a multilocus enzyme electrophoresis analysis of: (a) 60 strains of species which show lack of clonal structure due to common genetic exchange between strains, leading to randomization of alleles and few identical strains; and (b) 69 strains of a clonally structural species in which gene exchange is relatively uncommon leading to stable clonal types as indicated by the groups of numerous strains with identical electrophoretic patterns.

chromosome and linkage equilibrium is attained (Fig. 4.2). This is strong evidence for restricted chromosomal gene transfer in the environment among these clonal organisms. Some other species such as *Neisseria gonorrhoea* and *B. subtilis* have a low clonal structure and identical electrophoretic types are rarely encountered from the environment. This suggests that strains of these bacteria frequently transfer and recombine chromosomal DNA, leading to randomization of alleles. This is not particularly surprising given that both of these species are naturally transformable by exogenous DNA.

2. When sampling several thousand isolates by MLEE it has been revealed that the number of clones in a clonal species is fairly small with no more than a few hundred genotypes at most.

3. For pathogenic micro-organisms, the majority of cases of serious disease are associated with a small number of clonal types. Variance in pathogenicity is considerable within a species and some clones have obviously evolved increased virulence. In bacteria which infect several hosts there may be associations between host and clonal type. For example, *Bordetella bronchiseptica* clone ET-1 is generally recovered from swine, while clone ET-6 isolates are generally recovered from dogs.

The discovery of the clonal structure of bacterial populations is therefore having tremendous implications in epidemiology and in ecology where such studies can indicate the prevalence of chromosomal gene exchange in the environment. Interestingly, MLEE is increasingly being augmented by direct sequence evidence and in the case of *E. coli* the outcome was controversial suggesting that chromosomal exchange plays an important role in the evolution of this species (Dykhuizen and Green, 1991), whereas MLEE suggests it does not. No doubt the situation will be clarified as more data become available.

4.2   CELL ENVELOPE ANALYSES

Correlating with the different reactions to the crystal violet/iodine complex of Gram's stain, Gram-positive and Gram-negative organisms have very different cell envelopes. The Gram-positive bacterium is surrounded by a cytoplasmic membrane bounded by a thick layer of peptidoglycan, containing covalently bound teichoic acid. The mycobacteria and other acid-fast bacteria have a modified Gram-positive envelope in which considerable lipid material is intercolated into the peptidoglycan and they, therefore, stain Gram negative. True Gram-negative bacteria possess a double-layered envelope comprising the cytoplasmic membrane, a thin layer of peptidoglycan within a hydrophilic compartment, termed the periplasm, and an outer membrane comprising lipoprotein and lipopolysaccharide (Rogers, 1981). Structural variation in these various components and molecules can be used to classify and identify bacteria.

4.2.1 Peptidoglycan

Peptidoglycan (murein) is found in all bacteria, except mycoplasmas, planctomycetes and archaebacteria. It comprises amino-sugar backbones bearing tetrapeptide chains with a diamino acid in position three. These chains are cross-linked either directly or through an interpeptide bridge of L-amino acid residues between the diamino acid and the D-Ala residue at position four on an adjacent chain. The amino-sugar backbones are largely homogeneous, but the diamino acid of the tetrapeptide chain and the composition of the interpeptide bridge may show variation between taxa. This information is especially useful for Gram-positive bacteria; Gram-negative bacteria are, however, remarkably uniform in peptidoglycan structure.

Data from qualitative cell-wall sugar and amino acid analyses have been particularly valuable in the systematics of actinomycetes and related bacteria, which have been allocated to eight wall chemotypes. For example, the nature of the diamino acid proved particularly important for the classification of coryneform bacteria. *Corynebacterium diphtheriae* and related pathogenic strains have directly cross-linked *meso*-diaminopimelic acid-containing peptidoglycans (wall chemotype IV) like several pathogenic mycobacteria and nocardia, whereas other coryneform genera, such as *Arthrobacter* and *Microbacterium*, have the *meso*-diaminopimelic acid replaced by lysine and belong to wall chemotype VI (Goodfellow and Cross, 1984).

Methods for determining wall chemotype are fairly simple and rapid, being based on qualitative chromatographic analysis of acid hydrolysates of whole organisms or purified cell walls. More precise taxonomic data can be obtained from the primary structure of the peptidoglycan, but the analytical methods involved are specialized and beyond the scope of the average microbiological laboratory.

Polysaccharides and teichoic acids have received scant attention as chemotaxonomic markers.

4.2.2 Lipids

Lipids provide a wealth of taxonomic information that may be used for both classification and identification. There are four important classes: long-chain fatty acids; mycolic acids; polar lipids; and isoprenoid quinones.

1. Long-chain fatty acids. These can be released from polar lipids of the plasma membrane or the lipopolysaccharide of Gram-negative

Table 4.1   Some representative lipids from bacteria

(a) **Fatty acids**

n: Straight-chain acid
CH_3—$(CH_2)_{14}$—COOH

n-Hexadecanoic acid $(C_{16:0})$
(Palmitic acid)

i: Iso-branched acid
CH_3—CH—$(CH_2)_{12}$—COOH
      |
      CH_3

14-Methyl pentadecanoic acid
$(i\text{-}C_{16:0})$

a: Anteiso-branched acid
CH_3—CH_2—CH—$(CH_2)_{12}$—COOH
            |
            CH_3

12-Methyl tetradecanoic acid
$(a\text{-}C_{15:0})$

n:1: Monounsaturated acid
CH_3—$(CH_2)_5$—CH=CH—$(CH_2)_9$—COOH

11-Octadecanoic acid $(\Delta^{11}\text{-}C_{18:1})$
(*cis*-Vaccenic acid)

Δ: Cyclopropane acid

$\underset{CH_2}{\overset{\diagup\diagdown}{}}$
CH_3—$(CH_2)_5$—CH—CH—$(CH_2)_9$—COOH

11, 12-Methyleneoctadecanoic acid
$(\Delta C_{19:0})$
Lactobacillic acid

10-Methyl branched acid
                CH_3
                |
CH_3—$(CH_2)_7$—CH—CH_2—$(CH_2)_7$—COOH

10-Methyloctadecanoic acid
(10-Me $C_{19:0}$)
(Tuberculostearic acid)

ω-Cyclohexyl acid
$\diagup CH_2$—$CH_2\diagdown$
CH_2 $\quad$ CH—$(CH_2)_{10}$—COOH
$\diagdown CH_2$—$CH_2\diagup$

ω-Cyclohexylundecanoic acid

3-Hydroxy acid
CH_3—$(CH_2)_{10}$—CH(OH)—CH_2—COOH

3-Hydroxytetradecanoic acid
(3-OH $C_{14:0}$)

2- Hydroxy acid
CH_3—$(CH_2)_9$—CH(OH)—COOH

2-Hydroxydodecanoic acid
(2-OH $C_{12:0}$)

(b) **A typical mycolic acid** (from *Mycobacterium tuberculosis*)

$$CH_3\text{—}(CH_2)_{17}\text{—}\underset{\underset{CH_3}{|}}{CH}\text{—}\overset{\overset{O}{\|}}{C}\text{—}(CH_2)_{17}\text{—}CH\text{—}CH\text{—}(CH_2)_{19}\text{—}CH\text{—}\underset{\underset{C_{24}H_{49}}{|}}{CH}\text{—}\overset{\overset{OH}{|}}{CH}\text{—}COOH$$

with CH$_2$ bridge between the two central CH groups.

continued...

Table 4.1 *continued*

(c) Some phospholipids found in bacteria cells

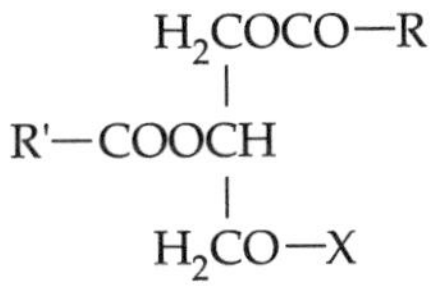

$$H_2COCO-R$$
$$R'-COOCH$$
$$H_2CO-X$$

Compounds	X

Phosphatidic acid — $-PO_3H_2$

Phosphatidylcholine —

$$-\overset{\overset{\textstyle O}{\|}}{\underset{\underset{\textstyle O^-}{|}}{P}}-OCH_2CH_2\overset{+}{N}(CH_3)_3$$

Phosphatidylethanolamine —

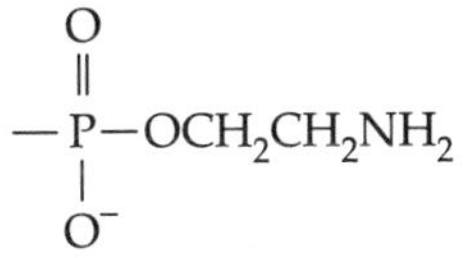

$$-\overset{\overset{\textstyle O}{\|}}{\underset{\underset{\textstyle O^-}{|}}{P}}-OCH_2CH_2NH_2$$

Phosphatidylserine —

$$-\overset{\overset{\textstyle O}{\|}}{\underset{\underset{\textstyle O^-}{|}}{P}}-OCH_2\underset{\underset{\textstyle NH_2}{|}}{CH}COO^-$$

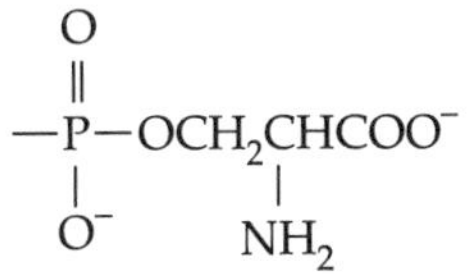

Phosphatidylinositol

$$-\overset{\overset{\textstyle O}{\|}}{\underset{\underset{\textstyle O^-}{|}}{P}}-O-$$

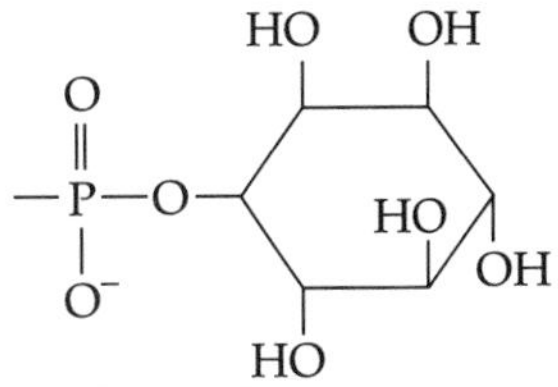

Phosphatidylinositol mannosides

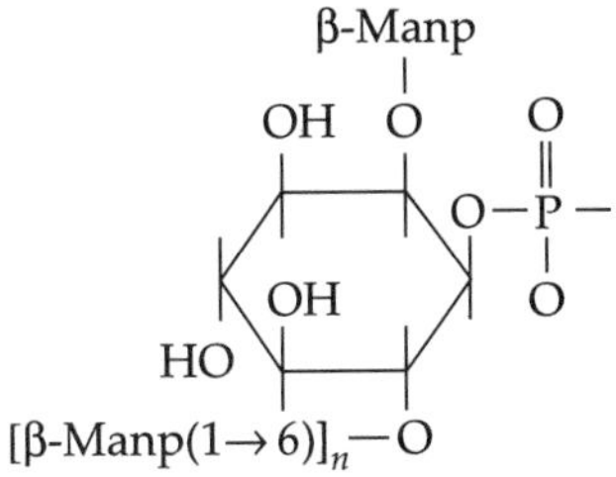

β-Manp

[β-Manp(1→6)]$_n$—O

continued...

Table 4.1   *continued*

(c) **Some phospholipids found in bacteria cells**

Phosphatidylglycerol

$$\begin{array}{c} CH_2-OPO_2^- \\ | \\ H-C-OH \\ | \\ CH_2OH \end{array}$$

(d) **Some bacterial isoprenoid quinones**

Naphthoquinones   Benzoquinones

Menaquinone   Ubiquinone

Demethylmenaquinone   Rhodoquinone

---

bacteria by esterification, and analysed by gas chromatography (GC). Variation in fatty acid composition includes the length of the carbon chain (which can range from about 8 to 26 carbons) the presence of branching, degree of unsaturation and the occurrence of ring structures (Table 4.1). The length of the alkyl chain influences membrane fluidity. Growth at higher temperatures promotes the synthesis of longer chains. Introduction of a double bond in the chain also affects membrane fluidity, and the degree of unsaturation is therefore of doubtful usefulness in taxonomy since it is so highly dependent on growth conditions. Hydroxylated fatty acids have the OH group in either position 2 or 3, and are found in almost all Gram-negative species (Table 4.2). Cyclopropane fatty acids are found in major quantities in both Gram-negative and Gram-positive bacteria. On the other hand, ω-cyclohexane and ω-cyclobutane fatty acids are so rare that they are a specific marker for the acidophilic, thermophilic endospore-forming bacteria previously typified by *Bacillus acidocaldarius* and now placed in the new genus *Alicyclobacillus* as *A. acidocaldarius*. Fatty acids, with branched alkyl chains, predominate in Gram-positive genera. Fatty acid patterns, generated by GC can be stored in a computer and analysed by the standard

Table 4.2 Some Gram-negative genera grouped according to their content of hydroxy-, methyl- and cyclopropane substituted fatty acids (after Jantzen and Bryn, 1985)

FA group	OH-FA	BR-FA	CYCLO-FA
Acinetobacter, Bordetella pertussis, Haemophilus, Moraxella, Neisseria, Pasteurella	+	–	–
Alcaligenes, Enterobacteriaceae, *Flavobacterium-Cytophaga 1*	+	–	+
Bacteroides, Flavobacterium-Cytophaga 2	+	+	–
Legionella	–	+	–
Brucella	–	–	+
Treponema, Gardnerella	–	–	–

Abbreviations: FA, fatty acid; OH-, hydroxylated; BR, branched (*iso* or *anteiso*); CYCLO, cyclopropane substituted.

numerical techniques of ordination and cluster analysis, to provide classifications. Moreover patterns from unknown organisms can be identified by comparison with those stored in the computer.

2. Mycolic acids. Some genera contain highly characteristic fatty acids which can be used as taxonomic markers of unique value. For example, the high molecular weight 2-alkyl-3-hydroxy-branched fatty acids (mycolic acids) are found only in the complex lipids of *Mycobacterium, Corynebacterium* and *Nocardia* strains (Komagata and Suzuki, 1987). These massive molecules have total carbon numbers from 24 to 90 and are responsible for the waxy, impermeable barrier found on these cells giving rise to the acid fast reaction and Gram-negative staining of these bacteria (Table 4.1). The lipids are located in the cell wall as free acids, glycolipids and complexed with polysaccharides.

3. Polar lipids. Bacterial membranes are largely composed of amphipathic polar lipids, which comprise a hydrophilic head group linked to two hydrophobic fatty acid chains (Table 4.1). Polar lipids are often referred to as free lipids, since they can be readily extracted by soaking cells in appropriate organic solvents. They are generally analysed by TLC, and often yield characteristic patterns. The most common polar lipids are phospholipids; other types include glycolipids and amino acid amides. Some polar lipids, for example, phosphatidylethanolamine, phosphatidylglycerol and diphosphatidylglycerol, are common and of little diagnostic value, but others, such as the phosphatidylinositol mannosides of actinomycetes are rarer, and of more diagnostic importance (Tables 4.1 and 4.3). A rapid presumptive test for archaea is the absence of esterified lipids and the presence of diether lipids.

4. Isoprenoid quinones. The isoprenoid or respiratory quinones are found in the plasma membranes of all aerobic bacteria and comprise

Table 4.3 Lipids of some actinomycetes with a wall chemotype IV (from Goodfellow and Cross, 1984)

Taxon	Long chain* fatty acids	No. of carbons	No. of double bonds	Predominant menaquinone[+]	Diagnostic phospholipids[++]
Corynebacterium	SU(T)	22–38	0–2	MK-8,9(H$_2$)	PI, PIM
Mycobacterium	SUT	60–90	1–2	MK-9(H$_2$)	PE, PI, PIM
Nocardia	SUT	46–60	0–3	MK-8(H$_4$)	
Saccharomonospora	SUIA	–	–	MK-9(H$_4$)	PE, PI, PIM

* S, straight chain; U, monounsaturated; T, tuberculostearic; I, *iso*; A, *anteiso*
[+] Abbreviations exemplified by: MK-8(H$_2$), menaquinone with two of the eight isoprene units hydrogenated
[++] PE, phosphatidylethanolamine; PI, phosphatidylinositol; PIM, phosphatidyl-mannosides Parentheses indicate not found on all strains

two general types; the isoprenoid menaquinones (2-methyl-3-polyprenyl-1-naphthoquinone, formerly known as vitamin K2) and isoprenoid ubiquinones (2,3-dimethoxy-5-methyl-6-polyprenyl-1,4-benzoquinone, or coenzyme Q). These are both large classes of molecules, in which the length of the polyprenyl side chain can vary from 1 to 14 isoprene units and also in the degree of saturation (Table 4.1). They are generally extracted with organic solvent and analysed by reverse-phase TLC and/or HPLC (Collins, 1985).

Most bacteria contain either menaquinones or ubiquinones, or both. For example, the aerobic Gram-negative rods generally contain only ubiquinones with members of *Pseudomonas* possessing ubiquinones with nine isoprene units, whereas several enteric genera contain mixtures of menaquinones and ubiquinones. Most members of *Bacillus* have menaquinones, and lactic acid bacteria generally lack isoprenoid quinones. Collins and Jones (1981) have surveyed the distribution of isoprenoid quinones among bacteria.

4.3   END-PRODUCTS OF METABOLISM

Traditional taxonomic tests, such as the methyl red test or Voges Proskauer reaction, determine the end-products of metabolism, but for some taxa a more specific analysis of these metabolites is useful. Quantitative analysis using GC or HPLC is particularly valuable and lends itself to computerized data processing for generating classifications and effecting identifications.

Fermentative bacteria usually yield more complex patterns of end-products than aerobic strains, and the technique is particularly valuable for anaerobic bacteria. Most methods concentrate on the acid end-products, particularly the monocarboxylic acids such as acetic, propionic and butyric.

4.4 COMPLETE CELLS

When bacteria are grown under standardized conditions and thermally degraded using a direct heating probe, laser, or heating foil at 300–400°C (although higher temperatures up to 700°C may be used) in an inert atmosphere (pyrolysed, Py), the products provide a 'finger print' of the cell. These chemical profiles are very complex but can be analysed by either GC or mass spectrometry (MS). Most of the earlier analytical work used GC, but problems with reproducibility led to the development of Py-MS which is faster, more reproducible and more easily automated than Py-GC . The details of Py-MS have been described by Gutteridge (1987); suffice to say that the output is a complex pattern of peaks reflecting the different products of pyrolysis. Since most bacteria tend to produce the same pattern of peaks, the quantitative data, represented by peak heights, which vary significantly and reproducibly between taxa, are used. The data are generally analysed using ordination procedures, since cluster analysis has not proved entirely successful, probably because of the complexity of the data. For this reason Py-MS is used to confirm existing classifications rather than generate new ones. It is ideally suited for identification purposes (see Chapter 7) since it is rapid (less than 10 min is required per sample), widely applicable and readily automated. Moreover, dedicated Py-MS machines are now being marketed which are finally reducing the prohibitive costs of the earlier machines and making this technology more widely available.

4.5 CONCLUSIONS

Chemical analyses have become a major influence in bacterial taxonomy as new methods and techniques are developed. They have proved particularly successful for classifying and identifying organisms in which morphological and physiological characters have been few or have led to confusing classifications through undue importance being placed on them. This is particularly relevant to the actinomycetes and related Gram-positive bacteria, the classification of which has been revolutionized by chemosystematics (Goodfellow, 1989). In many instances such as cell wall structures and lipid analyses, the chemosystematic classifications are consistent with phylogenetic classifications derived from 16S rRNA sequences and are invaluable in building up a complete picture of microbial relationships at the supra-generic level. This is essential for the development of comprehensive and stable classifications. At the other extreme, the application of MLEE is providing insight into the structure and evolution of the species itself. Finally, it must be emphasized that chemotaxonomic analyses, such as SDS-PAGE of cellular proteins, GC analysis of lipids and Py-MS, lend themselves to computerized data handling and automation and this technology is having a tremendous influence in rapid, automated identification, particularly of clinically important microbes.

REFERENCES

Collins, M.D. (1985). Isoprenoid quinone analysis in bacterial classification and identification, in *Chemical Methods in Bacterial Systematics* (M. Goodfellow and D.E. Minnikin, Eds.), pp. 267–288. Academic Press, London.

Collins, M.D. and Jones, D. (1981). Distribution of isoprenoid quinone structural types in bacteria and their taxonomic implications, *Microbiological Reviews*, **45**, 316–354.

Dykhuizen, D.E. and Green, L. (1991). Recombination in *Escherichia coli* and the definition of biological species, *Journal of Bacteriology*, **173**, 7257–7268.

Goodfellow, M. (1989). Suprageneric classification of the actinomycetes, in *Bergey's Manual of Systematic Bacteriology, Volume 4* (S.T. Williams, M.E. Sharpe and J.G. Holt, Eds.), pp. 2333–2339, Williams and Wilkins, Baltimore.

Goodfellow, M. and Cross, T. (1984). Classification, in *Biology of the Actinomycetes* (M. Goodfellow, M. Mordarski and S. T. Williams, Eds.), pp. 7–164, Academic Press, London.

Gutteridge, C.S. (1987). Characterization of microorganisms by pyrolysis mass spectrometry, *Methods in Microbiology*, **19**, 227–272.

Jackman, P.J.H. (1985). Bacterial taxonomy based on electrophoretic whole cell protein patterns, in *Chemical Methods in Bacterial Systematics* (M. Goodfellow and D.E. Minnikin, Eds.), pp. 115–130. Academic Press, London.

Jantzen, E. and Bryn, K. (1985). Whole-cell and lipopolysaccharide, fatty acids and sugars of Gram-negative bacteria, in *Chemical Methods in Bacterial Systematics* (M. Goodfellow and D.E. Minnikin, Eds.), pp. 145–172, Academic Press, London.

Kersters, K. (1985). Numerical methods in the classification of bacteria by protein electrophoresis, in *Computer-assisted Bacterial Systematics* (M. Goodfellow, D. Jones and F.G. Priest, Eds.), pp. 337–365, Academic Press, London.

Komagata, K. and Suzuki, K.-I. (1987). Lipid and cell wall analyses in bacterial systematics, *Methods in Microbiology*, **19**, 161–208.

May, B. (1992). Starch gel electrophoresis of allozymes, in *Molecular Genetic Analysis of Populations–A Practical Approach* (A.R. Hoelzel, Ed.), pp. 1–27, IRL Press, Oxford.

Rogers, H.J. (1981). *Bacterial Cell Structure*, Van Nostrand Reinhold, Wokingham.

Selander, R.K. and Musser, J.M. (1990). Population genetics of bacterial pathogenesis, in *The Bacteria, Volume XI* (B. Iglewski, Ed.), pp. 11–36, Academic Press, Orlando.

Selander, R.K., Caugant, D.A., Ochman, H. *et al.* (1986). Methods for multilocus enzyme electrophoresis for bacterial population genetics and systematics, *Applied and Environmental Microbiology*, **51**, 873–884.

CHAPTER 5

Phylogenetics

The construction of a tree depicting evolutionary pathways of all organisms, living and dead, has been a primary aim of many systematists. Those interested in the systematics of higher animals and plants have benefited from a fossil record of evolutionary events to guide them in their deliberations and to provide evidence to support their theories. Microbiologists have a dearth of such objective information. Therefore, a particularly exciting development was the realization that organisms have in their genome (and proteins) records of the changes that have occurred since divergence from a common ancestor (Zuckerkandl and Pauling, 1965). The extent and nature of the differences among nucleotide sequences provides an incisive insight into the phylogenetic relationships of all organisms; the more homogeneous the sequences, the more closely related the organisms and conversely, the more diverse the sequences the more distant the organisms. Moreover, since changes happen more or less randomly in time, the rate of change of nucleic acid sequences acts as a molecular chronometer and permits estimation of the elapsed time, in a relative sense, between evolutionary events; the so-called 'evolutionary clock' (see Woese, 1987). The study of evolutionary relationships using the techniques of molecular biology (molecular phylogeny) aims to represent the evolutionary pathways from the earliest organisms, about 4 billion (thousand million) years ago, to the great array of biodiversity we see today. In this chapter, we will examine the methodology of molecular phylogeny, describe some of its major findings and also highlight some of the controversies it is promulgating.

5.1 GENERAL CONSIDERATIONS

All life forms, both extinct and extant, share a common origin which can be traced back to a few organisms that inhabited this planet about 4 billion years ago. Consequently, all animals, plants and micro-organisms are related by descent to each other. Closely related organisms are descended from more recent common ancestors than are distantly related ones. It is generally assumed that evolution follows a pattern of successive branchings into populations in which evolutionary change subsequently proceeds independently. The aim of the phylogeneticist is to determine the pattern of this branching and represent it as a tree (cladogram) from ancestral forms through to the organisms as we see them today. These cladograms may have a time axis, an evolutionary distance axis or may simply record the branching of lineages. Before we deal with methods for generating cladograms, it is necessary to cover two general aspects of this discipline.

5.1.1 Parsimony

The concept of parsimony dates from Darwin and postulates that evolution proceeds along the shortest possible pathway with the fewest number of steps. This concept of minimal evolution is essential if unrealistic and bizarre theories of evolutionary change are to be avoided. Parsimony is the guiding concept used to construct minimal length or most parsimonious cladograms. Sneath (1983) suggests a less restricted definition of parsimony as 'preferring fewer assumptions or simpler explanations', which then includes maximal compatibility cladograms. These are based on compatibility with the largest number of characters, irrespective of the number of changes that need to be made and are collectively referred to as distance matrix methods. Minimal length and maximal compatibility cladograms are closely related mathematically.

5.1.2 Trees

Cladograms are displayed as branched diagrams or trees. A phylogenetic tree is a graph composed of nodes and branches in which only one branch connects the two adjacent nodes. The nodes of the tree represent the taxonomic units which may be species, individuals or genes. The external nodes represent extant organisms; internal nodes represent ancestral organisms. The tree can be configured in one of two ways. Unscaled trees (Fig. 5.1(a)) are similar to phenograms. The external nodes are aligned and the branches are drawn according to a scale, generally a time scale. In a scaled tree (Fig. 5.1(b)) the lengths of the branches are proportional to the number of molecular changes which have taken place. The trees may be rooted, that is have an origin that represents the common ancestor and from which the branches extend, or may be unrooted with no origin or direction (Fig. 5.2). A rooted tree presents a unique path leading to all nodes in which the direction of each path corresponds to evolutionary time. An unrooted tree presents the phylogenetic relationships but does not specify the evolutionary path.

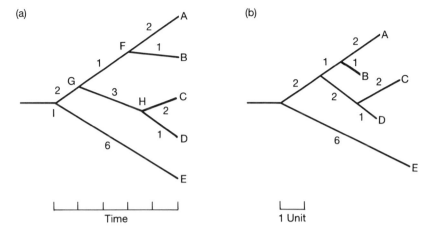

Fig. 5.1 Two alternative representations of a phylogenetic tree for five OTUs. (a) Unscaled branches: extant OTUs are lined up and nodes are positioned proportionally to times of divergence. (b) Scaled branches: lengths of branches are proportional to the numbers of molecular changes (from Li and Graur, 1991, with permission).

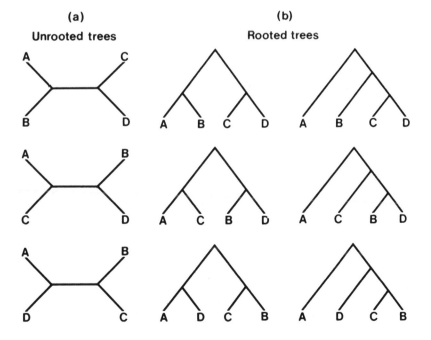

Fig. 5.2 (a) The three possible unrooted trees for four individuals. (b) Six of the 15 different rooted trees for four individuals.

Table 5.1   Possible numbers of rooted and unrooted trees for 1—10 OTUs

No. of OTUs	No. of rooted trees	No. of unrooted trees
2	1	1
3	3	1
4	15	3
5	105	15
6	954	105
7	10 395	954
8	135 135	10 395
9	2 027 025	135 135
10	34 459 425	2 027 025

Note that the number of unrooted trees for n OTUs is equal to the number of rooted trees for $n-1$ OTUs.

Whether the tree is rooted or not has considerable bearing on the computation involved. For four individuals, there are 15 different rooted trees but only three unrooted versions, and, for five individuals, the figures are 90 and 15, respectively (Fig. 5.2(b)). The numbers of possible rooted trees and unrooted trees for an increasing number of OTUs rapidly becomes very large (Table 5.1). For 10 OTUs, there are 35 million rooted trees and only one of these is the correct solution (i.e. is consistent with the data) and properly represents the speciation events that gave rise to the OTUs as we view them today. This is the 'true' tree which we pursue.

5.1.3   Algorithms versus optimality criteria

How do we find the 'true' tree among the millions of possibilities? In effect, we are trying to reconstruct history on incomplete evidence and searching for the best approximation to the actual events. There are essentially two approaches to phylogenetic inference. One approach aims to optimize the algorithm to provide the tree which most accurately represents the data (or from which the data could be best generated). Such algorithmic methods are fast but make no attempt to explain the evolutionary process directly. Moreover, there is no ranking of preferred trees. Methods such as cluster analysis fall into this category.

The second approach separates the algorithm for producing the tree from some measure of how accurately the data are reflected in the tree. An optimality criterion is employed to compare trees and choose the one which most accurately represents the 'true' tree. Thus the inference proceeds in two steps; first, determination of an optimality criterion for evaluating a given tree and comparing it to others, and second, to compute the ranking of trees based on this objective measure. Different optimality criteria have been suggested, such as parsimony, sum of squares of differences between tree and observed distances and others (see Penny *et al.*, 1992). Because numerous trees must be compared, these 'objective function' or optimality criterion methods are slower than algorithmic methods. We shall examine some of these approaches below.

5.2   DATA FOR PHYLOGENY

For many years, the only data available for inferring phylogeny were traditional phenotypic characters. For the animal and plant systematist, careful choice of character together with a wealth of fossil and embryological evidence allowed the construction of plausible cladograms, and culminated in the works of Hennig and his theory of monophyletic groups (see Chapter 1). With bacteria, phenotypic characters in the form of morphology, cell structure and comparative biochemistry and physiology have been available for some time, but they were of limited use for phylogenetic purposes because the choice of early (primitive) and derived characters was entirely subjective. There was (and still is) little useful fossil evidence with which to confirm ancestral phenotypes. Thus, theories of bacterial cladogeny were numerous and almost invariably contradictory. The morphological scheme of Kluyver and Van Niel in the 1930s assumed that the coccus is the simplest and, therefore, most primitive form, and the more complex structures, such as rods and mycelia, are derived from this. One line of descent resulted in the streptomycetes, another in the endospore-forming bacilli, and a third in the spirilla. By contrast, Lwoff and Knight, a decade later, suggested that bacteria evolved with a progressive loss of anabolic activity. Thus, the nutritionally less exacting bacteria are the most primitive and the more fastidious have evolved a specialized way of life more recently as they adopted symbiotic and parasitic lifestyles with higher animals and plants. Conversely, it may be argued that fastidious heterotrophs living in the 'primordial' soup, from which all life forms were derived, were the most primitive. It is, therefore, apparent that phenotypic features are of limited value for deriving a cladogeny for bacteria, since there is no evidence to favour any one of the above theories and there is negligible information about any selective pressures that might have operated in natural habitats over long periods of time.

The possibility of constructing a phylogeny of micro-organisms became more realistic with advances in molecular biology. Given a constant rate of evolutionary change, divergent evolutionary pathways and no gene transfer (but see sections 5.3.1 and 5.3.2), homologous sequences in DNA, RNA or protein must represent inherited sequences from ancestral forms. The amount of homology will reflect the phylogenetic relatedness; high homology implies recent divergence from a common ancestor; low homology implies early divergence from

Phylogenetics

a primitive ancestor. Therefore, current approaches to prokaryotic phylogeny use macromolecule sequences to construct cladograms consistent
with theories of rates of evolutionary change, divergence and convergence of pathways, and possibilities for lateral gene transfer.

The first sequences to be used in molecular phylogeny were those of
proteins. Cytochrome *c* and globulins are the best examples and were
important in showing a close correlation between cladograms derived
from the fossil record and those from molecular data. These studies
gave confidence to molecular phylogenetics but were not applied to
bacterial systematics to any great extent. Today, the ease with which
DNA can be sequenced compared with proteins has led to DNA
sequences comprising the vast bulk of the data for molecular phylogeny
reconstruction. Very often the sequences are translated into proteins
before analysis, however, to avoid problems arising from variation in
G+C content. But, we are still left with the choice of which gene
sequences to use, since DNA sequencing technology cannot yet provide
us with complete genome sequences for comparison. The choice tends

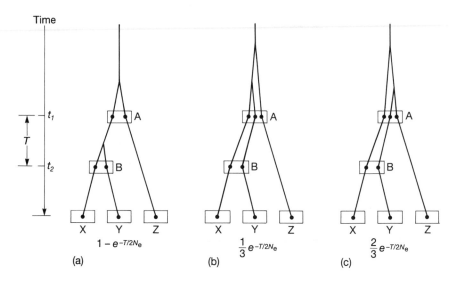

Fig. 5.3 Three possible relationships between a species tree (rectangles) and a gene
tree (dots). In (a) and (b), the topologies of the species trees are identical to those of
the gene trees. Note that in (a) the time of divergence between the genes is roughly
equal to the time of divergence between the populations. In (b), on the other hand,
the divergence between genes X and Y greatly predates the time of divergence
between the respective populations. The topology of the gene tree in (c) is different
from that of the species tree. For neutral alleles, the probability of occurrence t_1 is the
time at which the first speciation event occurred and t_2 is the time at which the
second speciation event occurred. $T = t_1 - t_2$, and N_e is the effective population size
(from Nei, 1987, with permission).

to be for strongly conserved genes; those which code for an RNA or protein with a highly specific function which therefore has strong constraints on it, limiting the amount of sequence variation it can accumulate. The rRNA genes, elongation factors involved in translation (Cammarano *et al.*, 1992), ATPase and glutamine synthetase all have their proponents (Pesole *et al.*, 1991). However, it could be argued that genes, that undergo little, if any, selection pressure because they do not lead to a functional product, such as the pseudogenes of eukaryotic cells, might accumulate mutations at a constant rate and therefore be better predictors of evolution. The answer to this dilemma is by no means clear, although it is known that pseudogenes evolve faster than genes under strong selection pressure. Another consideration is the distribution of the gene. rRNA genes are attractive because all organisms with the exception of viruses possess these sequences.

Most molecular phylogenetics, particularly of bacteria and archaebacteria have been derived from 16S rRNA sequences (Woese, 1987). It must be emphasized that trees derived from these sequences are 'gene' trees rather than 'species' trees and the two may differ. Most importantly, the branching pattern of a gene tree may be different from that of a species tree if the time of divergence between the genes greatly predates the time of divergence between the respective populations (Fig. 5.3). To avoid this type of error several genes should be used in the reconstruction of the phylogeny.

5.3   PHYLOGENIES FROM MACROMOLECULAR SEQUENCES

The process of deriving a cladogram from macromolecular data is essentially the same, irrespective of whether DNA or protein sequences are used. However, because nucleic acid sequences are the more common, we will deal exclusively with these here.

5.3.1   Alignment of sequences

The first stage is to align the sequences from the different sources such that the maximum amount of homology is obtained. Usually, this will involve inserting gaps into sequences to allow for additions or deletions of residues. The best possible alignment between two sequences is the one in which the number of mismatches and gaps is kept to a minimum. However, reducing the number of mismatches results in increasing the number of gaps and *vice versa*. For example, the mismatches in the two sequences X and Y shown below could be removed by the introduction of four gaps:

X: G C A – G C A A T
Y: G – A A G – A – T

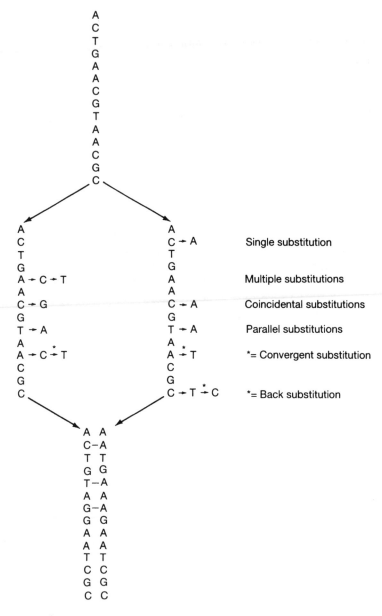

Fig. 5.4 Two homologous DNA sequences which descended from an ancestral sequence and accumulated mutations since their divergence from each other. Note that although 12 mutations have accumulated, differences can be detected at only three nucleotide sites. Note further that coincidental substitutions, parallel substitutions, convergent substitutions, and back substitutions all involve multiple substitutions at the same site, though they may be in different lineages (from Li and Graur, 1991, with permission).

Alternatively, the gaps could be minimalized to one and the mismatches maximized:

X: G C A G C A A T
Y: G A A G A T – –

Finally, we could compromise the mismatches to two and gaps to two:

X: G C A G C A A T
Y: G – A A G A – T

Sequence alignment algorithms apply gap 'penalties' to minimize the number of gaps while achieving the fewest number of mismatches. Common methods of calculating homology between aligned sequences are based on both similarity (e.g. the S_{AB} values of the cataloguing process) and distance measures and provide figures similar to those provided by measures of phenetic affinity.

5.3.2   Allowance for nucleotide substitution

For closely related (recently diverged) sequences, the number of mismatches between two sequences will probably reflect the degree of divergence reasonably accurately because there will have been little time for more than one substitution to have occurred at any one site. However, for distantly related sequences, the simple assessment of mismatches will underestimate the degree of divergence, because it will make no allowance for multiple substitutions and back mutations that are not detected by simple frequency counts. Some examples of undetectable variation in sequences are shown in Fig. 5.4. Various transformations based on theoretical and empirical premises have been proposed, initially for protein sequences. The equation of Margoliash and Smith (1965) calculates the expected number of changes:

$$D' = n \ln [n/(n-D)]$$

where there are D differences in n sites. For rRNA sequences, Hori and Osawa (1979) calculate the rate of nucleotide substitution:

$$K_{nuc} = -(3/4)\ln [1.0-(4/3)\lambda]$$

where λ is the fraction of different residues. These corrections give a higher number of estimated changes than the observed changes and attempt to account for unobserved sequence variation.

5.4   TREE RECONSTRUCTION

Once the relationship between the molecules has been calculated, the next stage is the construction of the tree. This is a very complicated subject only recently discussed in reviews in straightforward terms (Penny *et al.*, 1992; Swofford and Olsen, 1990). There are several approaches to tree reconstruction of which we shall examine just three of the more common: distance matrix, maximum parsimony and maximum likelihood methods.

5.4.1 Distance matrix methods

Distance matrix methods are algorithmic in their approach. They use the table of sequence differences without recourse to the characters or character states. Such methods are useful for DNA reassociation data or serological analyses of proteins, where actual sequences are not known. Such methods, for example UPGMA clustering (see Chapter 2), provide a phenogram and the result is heavily influenced by assumptions of evolutionary rates and divergence. If evolutionary rates are constant in different lineages and convergence is minimal the phenogram and cladogram will be congruent (an ultrametric) and UPGMA clustering will provide a reasonable approximation to the true cladogram. If rates of evolution are unequal, misleading results will be obtained (see section 5.5). Woese and his colleagues (Stackebrandt and Woese, 1982; Woese, 1987) have used UPGMA clustering of similarities of 16S rRNA catalogue sequences from prokaryotes extensively with impressive results; but the classification is essentially phenetic.

Optimality criteria can be used in connection with distance matrix techniques to determine the best tree. The neighbours relation method uses a distance matrix to provide a tree. In Fig. 5.5, OTUs A and B are neighbours as are C and D because they are connected through a single internal node. A and D or B and C are not neighbours by this definition. Following the simple example shown in Fig. 5.5, let dAB, dAC, dAD, dBC, dBD and dCD be the six determined evolutionary distances for four species A, B, C and D. If the data fit a tree topology the following two conditions should hold:

$$dAB + dCD < dAC + dBD$$

and

$$dAB + dCD < dAD + dBC$$

In the simple case of four OTUs, the above two conditions can be used to identify the neighbours (A and B; C and D) and the topology of the tree is thus determined. With more OTUs, all possible branching orders are examined and the one that best fits the data is indicated as the true tree. However, the total number of trees is very large, as we saw previously (Table 5.1) so algorithms combining UPGMA clustering to provide the framework of the tree are combined with neighbour

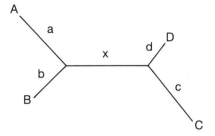

Fig. 5.5 Unrooted tree for four OTUs (A, B, C and D) and the distances (a, b, c, d, x) between them.

analysis to determine the best tree. Although this method is not overtly phylogenetic in the relationships determined, the trees produced are not intended to represent phenetic relationships.

5.4.2 Maximum parsimony methods

The governing assumption of maximum parsimony trees is that the true tree is the one that requires the fewest number of mutational changes to explain the differences observed between the gene sequences. This is the optimality criterion adopted. Maximum parsimony methods may also be described as 'character state' methods to distinguish them from those based on distance measures since they use the raw data rather than estimates of distance or similarity and the concept of 'informative sites'. A nucleotide site (a common base position in a set of sequences) is defined as informative if it favours only some of all the possible trees (see Fig. 5.6). For example, a constant base in all sequences is obviously not informative. Similarly, a site which is variable, but does not favour one tree over any others, is non-informative and not considered in the formulation of the tree.

To produce a maximum parsimony tree, all of the informative sites are first identified and the tree supported by the largest number of informative sites is the maximum parsimony tree. In other words, the minimum number of substitutions (changes) for each possible tree is determined, and the tree for which there is the fewest substitutions (minimum length) is calculated. Having determined the topology of the tree, the next stage is to estimate the branch lengths, a complicated process, but the end result is to scale the branches according to the number of site changes.

5.4.3 Maximum likelihood phylogenies

Given three entities, the data, a possible evolutionary tree and a model of evolutionary change, the probability of obtaining the observed data

					Site					
Sequence		1	2	3	4	5	6	7	8	9
	1	A	A	G	A	C	T	G	C	A
	2	A	G	C	C	C	T	G	C	G
	3	A	G	A	T	T	T	C	C	A
	4	A	G	A	G	T	T	C	C	G
						*		*		*

(a)

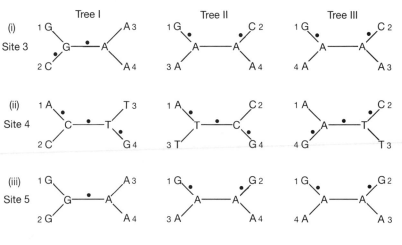

(b)

Fig. 5.6 Determination of informative sites for maximum parsimony methods of tree reconstruction. (a) Nucleotides occurring at nine sites in sequences from four OTUs. The informative sites are denoted by an asterisk. (b) Three possible trees for sites 3, 4 and 5 from the four sequences shown in (a). The terminal nodes are the nucleotides from homologous positions in the four sequences. Each dot indicates a substitution is inferred on that branch. Site 3 (top) is not informative since two changes are required in each of the three possible trees. Similarly, site 4 is not informative since three changes are required on each tree. Site 5, however, requires only one change in tree I, but two changes in trees II and III, and is therefore informative (from Li and Graur, 1991, with permission).

given the tree and model is computed. Since the model is invariant, the tree that maximizes this probability is the maximum likelihood estimate. It should be noted that this probability is not the probability that the tree is the correct one given the data and the model (Felsenstein, 1983).

The starting point is to develop a model of evolutionary change. Although this is potentially complex, relatively simple models are currently used. The Jukes–Cantor one-parameter model makes the assumptions that all four nucleotides are equally frequent, and that all substitutions are equally frequent, i.e. that transitions between two purines or two pyrimidines are as likely as transversions between a purine and pyrimidine or *vice versa*. The Kimura two-parameter model, on the other hand, sets independent rates for transitions and transversions which is a more realistic approximation to actual events. The mathematical expression of these models of evolutionary change quickly becomes complex and the interested reader is referred to specialist texts such as Li and Graur (1991) and for fuller details Swofford and Olsen (1990).

5.4.4 Rooting the tree

Algorithms for tree reconstruction usually generate unrooted trees because of the excessively large numbers of rooted trees. To place a root on a tree an outgroup is used. An outgroup is a fairly distant relative of the organisms under study (but not too distant) for which external evidence is available to document the earlier divergence of the group. For example, in Fig. 5.7, the shark sequence was used as the outgroup. The root is placed between the outgroup and the node connecting it to the other OTUs.

If no outgroup is available, a root can be placed on the assumption that the rate of evolution is approximately equal over all branches of the tree. The root is then placed at the midpoint of the longest pathway between two OTUs.

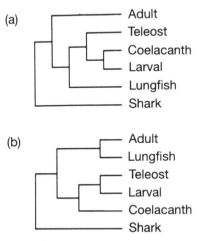

Fig. 5.7 Trees derived from ß-haemoglobin sequences using (a) UPGMA clustering and (b) a maximum parsimony procedure. 'Adult' and 'larval' refer to amphibian sequences. Note that the trees have little in common and that the maximum parsimony tree does not support a sister-group relationship between coelacanth and tadpole (adapted from Stock and Swofford, 1991).

5.5 CLADISTICS AND PHENETICS

Distance approaches, in particular UPGMA cluster analysis, are overtly phenetic and provide a classification which is phenetic in nature. If evolution has been constant and divergent this will be consistent with the phylogenetic classification. Character state approaches, such as maximum parsimony, which accommodate variations in evolutionary rates, are designed to give a more accurate representation of the true tree. But a reasonable question to ask is: do these approaches give different results?

A recent study of α- and β-haemoglobulin sequences from the coelacanth generated considerable correspondence in the journal *Nature* (see Scientific correspondence Vol. 353, pp. 217–219) and showed that the two methods do indeed give very different results. The initial study used UPGMA cluster analysis and from the tree shown in Fig. 5.7(a) claimed that the coelacanth is more closely related to tadpoles than is the lungfish. This result is at odds with morphological and other molecular data which place the lungfishes alongside the tetrapods (Gorr *et al.*, 1991). However, when the data were reanalysed using maximum parsimony, the relationships were changed and the lungfish was shown to be most closely related to the tetrapod (Fig. 5.7(b)). The reason for the failure of UPGMA to provide the most appropriate tree is due to unequal rates of evolution in the different lineages (Stock and Swofford, 1991). It is therefore important to remember that UPGMA-generated dendrograms of sequence data may generate misleading phylogenetic trees.

5.6 CONSTANCY OF EVOLUTIONARY RATES

If some branches evolve faster than others, then the deeper branching observed in some taxa may not reflect earlier divergence but simply a faster evolutionary 'clock'. To check if individuals within a group are isochronic (evolving at a similar rate), individuals within the group can be compared with an outgroup reference. If each member of the group shows a similar level of sequence homology with the reference, it may be argued that they have evolved at similar rates with regard to the reference and that the deeper branches reflect earlier divergence. When Stackebrandt and Woese (1982) used this approach to evaluate evolution rates derived from rRNA catalogues of bacilli, they showed that most bacilli appear to evolve at a constant rate with respect to *E. coli*.

When protein gene sequences are used for inferring phylogenies, it must be remembered that different functional classes of protein evolve at different rates. Histones seem to be among the most slowly evolving proteins with one substitution occurring in several hundred million years. However immunoglobulins evolve about 100 times faster. The difference in evolutionary rates is largely due to functional constraints on the molecules; indispensable proteins with conserved function have a lower probability that an amino acid substitution would be compatible with maintenance of function. Therefore they accumulate fewer mutations. At the other extreme, pseudogenes have no essential function and accumulate mutations very rapidly. When a time scale is added to a cladogram it is therefore important to compare data from different molecules.

If the absolute time is known for a branching event, it is possible to calculate the approximate times for other branches of the cladogram from nucleotide substitution rates (the molecular chronometer). Fossil evidence can be used to measure times of divergence, but the fossil record of early organisms (older than 100 million years) is poor. This, coupled with the uncertainty about rates of nucleotide substitutions in early organisms, complicates adding a time scale to cladograms of bacteria. Nevertheless, 5S rRNA sequence analysis suggests that Gram-positive and Gram-negative bacteria diverged about 1.2×10^9 years ago and geological evidence indicates that the photosynthetic bacteria were established $3.5–3.8 \times 10^9$ years ago, within a billion years of the earth's formation.

5.7 GENE TRANSFER

The prevalence of gene transfer in bacteria has led to the suggestion that the assumption of divergent branching in cladogram construction may not be acceptable and evolutionary lineages may frequently converge. Evidence for convergence through gene transfer can be obtained by searching for similar genes among organisms with very different backgrounds. Identification of such genes either represents extreme conservation or, more likely, recent transfer of the gene. Some antibiotic resistance genes are in this category, for example, the β-lactamases are immunologically related and have considerable DNA homology, whatever their origin. This suggests that the gene for β-lactamase has been transferred laterally among the bacterial kingdom. A recent analysis of 18 Class A β-lactamases showed the origin of the enzyme among the actinomycetes from which it migrated first into non-actinomycete Gram-positive lines such as *Bacillus*, and later into the Gram-negative bacteria (Kirby, 1992). The origin of β-lactamase is probably the penicillin-sensitive D-alanyl-D-alanine peptidase from actinomycetes which is involved in the synthesis of peptidoglycan and is the target for β-lactam antibiotics. Interestingly, the β-lactam production pathway of the fungi is a relatively recent addition to their genome from the actinomycetes by horizontal gene transfer (Miller and Ingolia, 1989).

The location of the β-lactamase gene within a transposable DNA sequence (transposon) which can integrate and excise from bacterial, plasmid or phage chromosomes, and in so doing replicate itself, largely explains the rapid lateral transfer of this gene (see Brown, 1992). Numerous phenotypic traits have been associated with plasmids and often transposons, including hydrocarbon and sugar catabolism, pathogenicity in plants and animals, heavy metal tolerance, and, of course, antibiotic resistance, suggesting that gene transfer could be an important feature in prokaryotic evolution.

The contribution of gene transfer to bacterial evolution can be estimated by comparison of cladograms based on different data. Two such cladograms will be incongruent if gene transfer were prevalent since it would be unlikely that patterns of transfer for different genes would be the same, i.e. the gene trees would be significantly different from the species tree. Although studies in this area are few, the phylogeny of the purple non-sulphur photosynthetic bacteria has been deduced from amino-acid sequences of cytochrome *c* and 16S rRNA cataloguing. The resultant cladograms were similar suggesting that transfer of cytochrome *c* genes between these bacteria was improbable (Ambler, 1985). Thus, although gene transfer of antibiotic resistance and individual enzymes such as alkaline phosphatase or α-amylase seem to be prevalent (Smith *et al.*, 1991), genes for the predominant physiological functions of bacteria do not seem to undergo rampant interspecies transfer (Woese, 1987).

5.8   BACTERIAL EVOLUTION

The remarkable achievements of molecular phylogeny have led to some fascinating, and controversial, findings with regard to bacterial evolution. Here we will examine just three aspects of this topic.

5.8.1   The Archaea

Prior to the advent of molecular sequence analysis, all bacteria were assigned to one large group: the prokaryotes. The cellular structure of the prokaryotes was considered sufficiently different from that of the eukaryotes to warrant separate 'kingdom' status. However, comparative sequencing of the 16S rRNA from some extremophilic bacteria, notably some thermophiles and halophiles, revealed that their rRNAs were so dissimilar from typical bacterial sequences, that the organisms should be placed in a separate phylum (Figs. 5.8 and 5.9). The sequence comparisons showed that there were not two primary lines of evolutionary descent, the prokaryotes and eukaryotes, but three: the eukaryotes, prokaryotes, and one represented by the archaebacteria, of

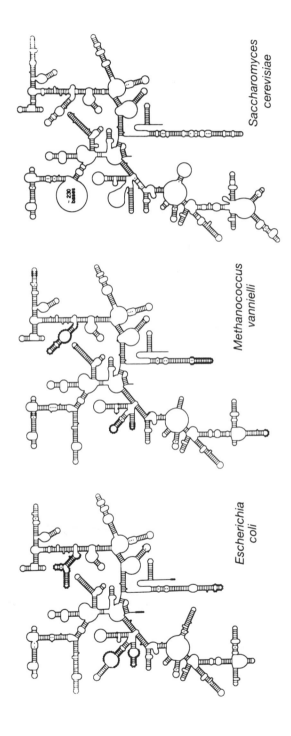

Fig. 5.8 Representative small-subunit rRNA molecules for the three 'domains', Bacteria (*Escherichia coli*), Archaea (*Methanococcus vannielli*) and Eucarya (*Saccharomyces cerevisiae*). Dots in the *E. coli* and *M. vannielli* sequences indicate areas where the two sequences differ most (from Woese, 1987, with permission).

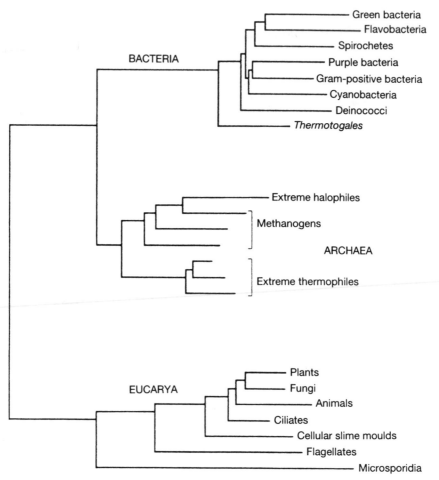

Fig. 5.9 A universal phylogenetic tree produced from an evolutionary distance matrix derived from an alignment of small subunit rRNA sequences (from Woese, 1992, with permission).

which the last two mentioned are prokaryotic in cellular structure. An alternative terminology (adopted here) refers to these phyla as the 'domains' Eucarya, Bacteria and Archaea (Woese *et al.*, 1990).

It is clear that many fundamental differences exist between the archaea and bacteria (Table 5.2). Each has its own characteristic rRNA, the two types are almost as different from each other as they are from eukaryotic rRNA. Archaeal RNA polymerases are of at least two types, and are structurally more complex than the single type found in Bacteria. Certain aspects of the protein synthesizing machinery are different in the three domains. Although the ribosomes of the Archaea and Bacteria are the same size, several steps in archaeal protein synthesis strongly resemble

Table 5.2　Summary of major differences between Archaea, Bacteria and Eucarya

Characteristic	Archaea	Eucarya	Bacteria
Membrane-bound nucleus	Absent	Present	Absent
Cell wall muramic acid	Absent	Absent	Present
Membrane lipids	Ether-linked	Ester-linked	Ester-linked
Ribosomes	70S	80S	70S
Initiator tRNA	Methionine	Methionine	Formylmethionine
Introns in tRNA genes	Yes	Yes	No
Ribosome sensitivity to diphtheria toxin	Yes	Yes	No
RNA polymerases	Several	Several	One
Sensitivity to:			
Chloramphenicol	No	No	Yes
Kanamycin	No	No	Yes
Streptomycin	No	No	Yes
Rifampicin	No	No	Yes

those in Eucarya (Table 5.2). Indeed, most antibiotics that inhibit transcription and translation in bacteria have no effect in Archaea. Moreover, the tRNAs of Archaea show a characteristic pattern of modified bases and the occurrence of introns in the tRNA genes is reminiscent of Eucarya. One of the most distinctive features of the Archaea is the chemistry of the membrane lipids. These are branched-chain, ether-linked lipids, common and unique to all Archaea. In some species the glycerol ethers are covalently joined to produce diglycerol tetraethers (Fig. 5.10).

The Archaea form two distinct lineages (Woese *et al.*, 1990). One kingdom, Chrenarchaeota includes the extreme thermophiles and the other kingdom Euryarchaeota, the methanogens and relatives (Fig. 5.9). The extreme thermophiles all grow anaerobically and most require sulphur as an energy source. Some will also grow aerobically, and not

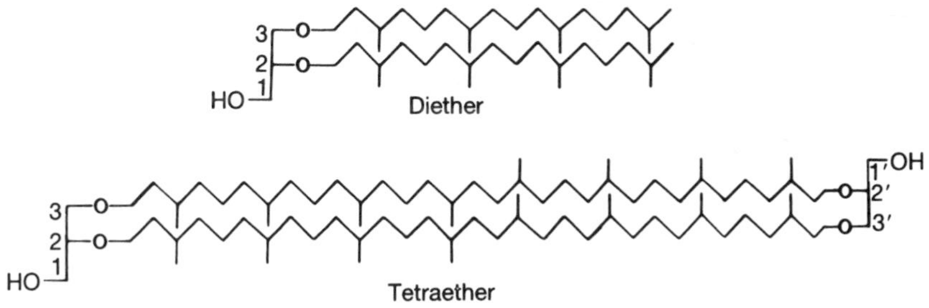

Fig. 5.10　Glycerolipids of archaebacteria; the phytanylglycerol diether and the dibiphytanylglycerol tetraether.

all are dependent on sulphur. These bacteria grow at extremely high temperatures with optima above 80°C. *Pyrodictium* holds the current record with a temperature optimum of approximately 106°C. The elemental sulphur is used as an electron acceptor under anaerobic conditions with various organic compounds acting as energy sources. However, one sulphur-dependent archaebacterium, *Sulfolobus* uses sulphur as an electron donor with oxygen or ferric ions as the electron acceptor. The intimate role of sulphur in the metabolism of these bacteria, together with their thermophilic character has led to the suggestion that they represent an ancient metabolic format.

The Euryarchaeota are more disparate in the phenotypes represented. The three methanogenic phyla based on *Methanococcus*, *Methanobacter* and *Methanomicrobium* show a wide variety of morphologies and habitats, but they are homogeneous with respect to their metabolism. They all share the unique ability to synthesize methane with various C_1 compounds such as CO_2, formate, methylamine, methanol or acetate as substrates. Methanogenesis from each substrate proceeds through a common step utilizing the unique coenzyme M and the complex enzyme system methyl-coenzyme M-methyl reductase.

The extreme halophiles comprise a highly distinctive group of organisms. They are generally obligate aerobes and grow in near saturated sodium chloride solutions. They have a very high internal potassium concentration (approaching 5 molar) and their mechanism of photosynthesis, based on bacteriorhodopsin, is unique.

The evolution of the Archaea probably began as an anaerobic, thermophilic, sulphur-metabolizing phenotype that remains extant. From this, methanogenic metabolism arose, from which halophilic methanogens evolved to become the aerobic, non-methanogenic extreme halophiles we see today (Woese, 1987).

The recognition of the Archaea as a separate, independent lineage has been accepted by the biological community, but controversy has been generated with regard to the rank of the taxon. Woese *et al.* (1990) consider the archaea to be so distinctive with regard to molecular structure of their rRNA genes that they should be awarded equal status with the existing kingdoms of prokaryotes and eukaryotes. To avoid preconceived notions attached to existing names, they suggested that three 'domains' of life should be recognized based on small subunit RNA characteristics termed Archaea, Bacteria and Eucarya (Woese *et al.*, 1990; Winker and Woese, 1991). Each of these domains would contain two or more kingdoms; for example the Eucarya would comprise the Animalia, Plantae, Fungi etc. and the Archaea would contain the Euryarchaeota and Crenarchaeota. Kingdoms in the Bacteria would be based on the classification described in Chapter 3. This terminology has been used in this book largely because we are dealing with micro-organisms and it is convenient to follow the Woese scheme, but understandably this proposal has created considerable dismay among many biologists. The Woese scheme was promoted to replace the five kingdom classification of Whittaker namely; Animalia, Plantae, Fungi, Protista (protozoa, algae, slime moulds etc.) and Procaryotae (bacteria). The proponents of the latter system (Mayr, 1991; Margulis and Guerrero, 1991) argue that to give the two subdivisions of the prokaryotes equal rank to the combined eukaryotic kingdoms is not commensurate with the enormous phenetic differences between the eukaryotic and prokaryotic cell. We return again to the dichotomy between the phylogenetic and phenetic classifications and which is to be preferred; one based on evolution, so far as it can be inferred, or one which takes into account the phenetic dimensions of the organisms. Whether the two classifications can be unified in this case remains to be seen.

5.8.2 The evolution of the eukaryotic cell

In the early 1970s, two possible routes to the derivation of the eukaryotic cell were envisaged; the 'autogenous' origin invoked progressive changes to the ancestral cell with continuing sophistication until the complex eukaryotic cell was formed. The alternative 'serial endosymbiosis' suggested that eukaryotic cells are chimeras assembled from several independent prokaryotic lines of descent. Mitochondria were derived from aerobic bacteria, plastids from photosynthetic bacteria and organelles of motility from spirochaetes (Margulis, 1981; Doolittle, 1988). Molecular phylogeny was able to provide unambiguous evidence for the latter route to the eukaryotic cell, a principle which is now well established. Gene sequences for 16S rRNA confirmed the cyanobacterial origin of the plastid and, although mitochondrial genomes show rapid rates of nucleotide substitution, sequence analysis indicated an ancestor from the α subdivision of the proteobacteria as the probable progenitor of the mitochondrion (Doolittle, 1988).

The most interesting question now is whether these organelles are monophyletic or polyphyletic and, if the latter, from how many ancestral lines they derive. The cryptomonad phytoflagellate *Cryptomonas* contains a structure termed a nucleomorph which is like a membrane-bounded nucleus associated with the chloroplast of the organism. Sequence studies of the 18S rRNA of the nucleomorph revealed that it is phylogenetically unrelated to the nucleus of the flagellate and closely related to sequences from the red algae, whereas the nuclear rRNA is most closely related to fungi and green plants. The red algae sequences form a clade, quite separate from the one that includes cyanobacteria and green plant plastids, indicating two distinct endosymbiotic routes and suggesting that red plastids have been secondarily acquired by cryptomonads (Douglas and Turner, 1991). Proponents of monophyly of plastids have argued that endosymbiosis would be a rare event requiring several improbable steps. However, others argue equally persuasively that in the early biosphere photosynthetic organisms were plentiful (perhaps 10^{20} organisms) each dividing 10 times per year. After one billion years this would have resulted in 10^{29} divisions and the improbable event then becomes quite likely (Penny and O'Kelley, 1991).

Another interesting question, with regard to the endosymbiont origin of the eukaryotic cell, is the nature of the host. Sequence analysis of 18S rRNA genes shows quite clearly that there is no extant lineage to which eukaryotic nuclei are particularly close. The eukaryotic line is ancient and diverged from prokaryotes before any internal divergences within the prokaryotes. Thus the original host need not have been like the eubacteria we study today and we are probably being misled when we consider the original 'protoeukaryote' as a cyanobacterium, mycoplasma or actinomycete.

5.8.3 The universal ancestor

The existence of 3.5-billion-year-old stromatolites (fossilized microbial mats consisting of layers of filamentous prokaryotes) implies the existence of photosynthetic bacteria such as *Chlorflexus* at that time. Working back, the thermophilic archaeal phenotype is consistent with the conditions on the early earth (high temperature; reducing anoxic atmosphere) making it likely that the Archaea arose within the first billion years of the earth's history when the temperature was high.

Woese (1987) argues for a progenote that had a simple genetic structure but not a genomic organization in which genes are arranged in a chromosome. Genes would be disjointed, perhaps in high copy number and probably based on RNA. The transition of the universal ancestor to ancestors of each of the primary lines of descent must have been rapid, taking less than 1 billion years, because stromatolites date photosynthetic bacteria to 3.5 billion years ago. It would seem that evolution at this stage was very rapid and it is likely that the universal ancestor was unlike any of its three major descendants.

In conclusion, molecular studies are providing a phylogenetic classification of the bacteria and higher organisms that is presenting the pheneticist and microbial biochemist with some fascinating challenges. Can we find traits that correlate with the phylogeny and finally marry the two systems of classification? The origins of the eukaryotic plastids, mitochondrion and flagellum also present interesting problems but the most important question relates to the nature of the earliest forms of life on this planet and it is to be hoped that molecular phylogeny will give us a glimpse of these, our earliest ancestors.

REFERENCES

Ambler, R.P. (1985). Protein sequencing and taxonomy, in *Computer Assisted Bacterial Taxonomy* (M. Goodfellow, D. Jones and F.G. Priest, Eds.), pp. 307–335, Academic Press, London.

Brown, T.A. (1992). *Genetics; A Molecular Approach* (2nd edn). Chapman & Hall, London.

Cammarano, P., Palm, P., Creti, R. *et al.* (1992). Early evolutionary relationships among known life forms inferred from elongation factor EF-2/EF-G sequences: phylogenetic coherence and structure of the archael domain, *Journal of Molecular Evolution,* **34**, 396–405.

Doolittle, W.F. (1988). Bacterial evolution, *Canadian Journal of Microbiology,* **34**, 547–551.

Douglas, S.E. and Turner, S. (1991). Molecular evidence for the origin of plastids from a cyanobacterium-like ancestor, *Journal of Molecular Evolution,* **33**, 267–273.

Felsenstein, I. (1983). Methods for inferring phylogenies: a statistical view, in *Numerical Taxonomy* (J. Felsenstein, Ed.), pp. 315–334, Springer-Verlag, Berlin.

Gorr, T., Kleinschmidt, T. and Fricke, H. (1991). Close tetrapod relationships of the coelocanth *Latimeria* indicated by haemoglobin sequences, *Nature,* **351**, 394–397.

Hori, H. and Osawa, S. (1979). Evolutionary change in 5S RNA secondary structure and a phylogenetic tree of 54 5S RNA species, *Proceedings of the National Academy of Sciences of the United States of America,* **76**, 381–385.

Kirby, R. (1992). Evolutionary origins of the class A and class B β lactamases, *Journal of Molecular Evolution,* **34**, 345–350.

Li, W.-H. and Graur, D. (1991). *Fundamentals of Molecular Evolution,* Sinauer Associates, Sunderland, MA.

Margoliash, E. and Smith, E.L. (1965). Structural and functional aspects of cytochrome *c* in relation to evolution, in *Evolving Genes and Proteins* (V. Bryson and H.I. Vogel, Eds.), pp. 221–242, Academic Press, New York.

Margulis, L. (1981). *Symbiosis in Cell Evolution,* W.H. Freeman and Co., San Francisco.

Margulis, L. and Guerrero, R. (1991). Kingdoms in turmoil, *New Scientist,* 23 March, pp. 46–50.

Mayr, E. (1991). Scientific correspondence, *Nature,* **353**, 122.

Miller, J.R. and Ignolia, T.D. (1989). Cloning and characterization of the beta-lactam biosynthesis genes, *Molecular Microbiology,* **3**, 689–695.

Nei, M. (1987). *Molecular Evolutionary Biology,* Columbia University Press, New York.

Penny, D. and O'Kelly, C.J. (1991). Seeds of a universal tree, *Nature,* **350**, 106–107.

Penny, D., Hendy, M. and Steel, M.A. (1992). Progress with methods for constructing evolutionary trees, *Trends in Evolution and Ecology,* **7**, 73–79.

Pesole, G. Bozetti, M.P. Preparata, G. and Saccone, C. (1991). Glutamine synthetase gene evolution; a good molecular clock, *Proceedings of the National Academy of Sciences of the United States of America,* **88**, 522–526.

Smith, J.M., Dowson, C.G. and Spratt, B.G. (1991). Localized sex in bacteria, *Nature,* **349**, 29–31.

Sneath, P.H.A. (1983). Philosophy and method in biological classification, in *Numerical Taxonomy* (I. Felsenstein, Ed.), pp. 22–37, Springer-Verlag, Berlin.

Stackebrandt, E. and Woese, C.R. (1982). The evolution of prokaryotes, *Symposium of the Society for General Microbiology,* **32**, 1–31.

Phylogenetics

Stock, D.W. and Swofford, D.L. (1991). Scientific correspondence, *Nature*, **353** 217–218.

Swofford, D.L. and Olsen, G.J. (1990). Phylogeny reconstruction, in *Molecular Systematics* (D.M. Hillis and C. Moritz, Ed.), pp. 411–501, Sinauer Associates, Sunderland, MA.

Winker, S, and Woese, C.R. (1991). A definition of the domains Archaea, Bacteria and Eucarya in terms of small subunit ribosomal RNA characteristics, *Systematic and Applied Microbiology*, **14**, 305–309.

Woese, C.R. (1987). Bacterial evolution, *Microbiological Reviews*, **51**, 221–271.

Woese, C.R. (1992). Prokaryote systematics: the evolution of a science, in *The Prokaryotes*, 2nd Edition (A. Balows, H.G. Trüper, M. Dworkin, W. Harder and K.M. Schleifer, Eds.), pp. 1–8, Springer-Verlag, New York.

Woese, C.R., Kandler, O. and Wheelis, M.L. (1990). Towards a natural system of organisms: proposal for the domains Archaea, Bacteria and Eucarya, *Proceedings of the National Academy of Sciences of the United States of America*, **87**, 4576–4579.

Zuckerkandl, E. and Pauling, L. (1965). Molecules as documents of evolutionary history, *Journal of Theoretical Biology*, **8**, 357–366.

CHAPTER 6

Nomenclature

What is in a name? This is perhaps a rhetorical, but nevertheless a necessary question. Names are labels that should convey a message about the organism. To mention the surname of a human being will conjure up a mental image of a group of related individuals. The use of the Christian name identifies an individual within that group. Confusion results if the same Christian name is applied to two or more human beings of the same family group. Similarly within bacteriology, it would be impossible for communication if two taxa possessed the same name. Names are not designed to characterize taxa, but represent a means for effective communication by labelling the entities. These labels are, of course, man-made, and constitute the procedure referred to as nomenclature (Wayne, 1991). However, the naming of bacterial groups evokes the best, and sometimes the worst, attributes of a scientist. The naming of new bacterial taxa is regarded by some scientists as a prestigious race to obtain immortality in the realms of biology. Where haste is evident, the outcome may be disastrous for reputations and bacterial taxonomy. The message is that it is often easier to allocate new names than do all the essential comparative work. However, it seems ironic that some of the noted bacteriologists are immortalized in the genus names of notorious pathogens. For example, Pasteur, Shiga and Yersin are remembered by the genera *Pasteurella, Shigella* and *Yersinia,* respectively.

So, the major function of nomenclature is communication. Consequently, it is important that names are readily pronounceable, sufficiently distinct from each other to avoid confusion, and stable. Bacteria may have common names, such as anthrax bacilli, and the more important binomial name *Bacillus anthracis.* This system owes its origin to the distinguished eighteenth century Swedish naturalist, Linnaeus. Thus species (see section 6.1) names comprise two words; the first representing the genus name, e.g. *Escherichia,* and the second consisting of the species identity (specific epithet), e.g. *coli.* In this example, the binomial name is *Escherichia coli.* Convention dictates that species names are italicized, as with the above example. These names are subjected to rules, as discussed in the International Code of Nomenclature of Bacteria (Sneath, 1992). A primary function is to ensure uniformity across international boundaries. After all, confusion would reign if the same organism was associated with a multiplicity of names depending upon the whims of scientists from different countries.

Initially, the procedures of bacterial nomenclature copied the system used by either the Botanical or Zoological codes. The special needs for bacterial nomenclature were initially mooted at the First International Congress of Microbiology in Paris during 1930. As a result, a Commission on Nomenclature and Taxonomy was formed, with a remit to investigate the problems and make recommendations to the Congress. The resultant resolutions highlighted the problems of bacterial nomenclature, but agreed that, wherever possible, the naming of micro-organisms should follow the existing Botanical and Zoological codes. However, it was also recommended that a representative committee, which would include representatives designated by the Botanical Congress, should be established with a view to examining the subject of bacterial nomenclature. This committee was named the Nomenclature Committee of the International Society for Microbiology. Its duties appertained to resolving problems in, and defining criteria to be used in, bacterial nomenclature.

At the Second International Congress for Microbiology, in London during 1936, it was agreed to formulate a code for bacteriological nomenclature, which was discussed at the Third International Congress of Microbiology held in New York in 1939. This Congress approved both the development of a formal code, and the establishment of a Judicial Commission. The Commission's terms of reference included the issuing of formal nomenclatural 'Opinions' upon request; the development of recommendations for emendation of rules of nomenclature; the preparation of lists of the names of microbial genera, and the publication of the International Rules of Bacteriological Nomenclature. The Proposed Bacteriological Code of Nomenclature was considered at the Fourth International Microbiological Congress in Copenhagen, 1947. This was duly revised and published (in English) in the *Journal of Bacteriology* and the *Journal of General Microbiology*. Rapid progress ensued, and at the next Microbiological Congress (Rio de Janeiro, 1950), publication of an 'official' journal, the *International Bulletin of Bacteriological Nomenclature and Taxonomy*, was authorized. Thereafter, the Bulletin grew into the much respected *International Journal of Systematic Bacteriology* (IJSB). As for the code, emendations and revisions occurred until the present form, with its legalistic connotations, i.e. the International Code of Nomenclature of Bacteria, was published in 1975 and in revised form in 1992. Alterations may only be made by the International Committee of Systematic Bacteriology at one of its plenary sessions.

Essentially, the Code seeks to ensure the presence of stable, clear meaningful names, which are derived from Latin or Greek, or if necessary, latinized. Of course, the name, unique to one taxon, must be validly described in the scientific literature, with the designation of a reference (type) strain. 'Priority' is given to the first valid publication of a name. However for a name to be valid, it must be initially published in the IJSB, or, if published elsewhere, subsequently listed in the IJSB. Moreover, names should not be changed without good taxonomic reasons. Thus, the Code should have a stabilizing effect upon nomenclature.

The Code considers taxonomic ranks from the level of subspecies up to and including class. These ranks include subspecies, species, subgenus, genus, subtribe, tribe, subfamily, family, suborder, order, subclass and class. The use of the term 'variety' is not encouraged, as this term is regarded as synonymous with subspecies. However, 'biovar', 'serovar' and 'pathovar' are accepted, although they are not covered by the Code. It should be emphasized that taxonomic ranking above family level is generally unclear in the case of bacteria.

The formation of names may be a difficult task, because consideration has to be given to grammar. Perhaps, the most straightforward aspect is the formulation of genus names, which appear to be constructed in an arbitrary fashion. With this case, the name may be derived from any word, although it is desirable that the name is descriptive or, at least, taken from the surname of a person associated with the organism, in particular, or bacteriology in general. Names should not be used which are already established in botanical or zoological taxonomy. Whatever the choice, the name is feminized, and always written in italics or underlined. The species name (or epithet) is similarly italicized, and, as a general rule, it may not be hyphenated, and has to agree in gender with the genus name. On no account shall a specific epithet be used more than once to describe two or more species within the same genus. The specific epithet should convey some meaningful message about the organisms, such as habitat or a distinctive characteristic. Alternatively, there may be a wish to name a species after a person, preferably the scientist connected with the isolation and/or characterization of the organism. This is permissible within the rules of the Code. Similar criteria apply to the delineation of subspecific names.

It may be observed that within the scientific literature, it is accepted convention to write the species name in full initially, e.g. *Pseudomonas aeruginosa*, but the genus name may be subsequently abbreviated to one letter followed by a full stop, e.g. *P. aeruginosa*. This is satisfactory providing that confusion does not result, such as may happen if reference is made to two or more genera beginning with the same letter, e.g. *Acinetobacter* and *Actinomyces*. With these cases, the genus name should be spelt out in full or abbreviated to a two or three letter code, such as *Aci.* and *Act.* for *Acinetobacter* and *Actinomyces*, respectively.

Table 6.1　Formation of bacterial names up to and including Order (based on Lapage *et al.*, 1975)

Taxonomic rank	Suffix	Example
Order	-ales	Pseudomonadales
Suborder	-ineae	Pseudomonadineae
Family	-aceae	Pseudomonadaceae
Subfamily	-oideae	Pseudomonadoideae
Tribe	-eae	Pseudomonadeae
Subtribe	-inae	Pseudomonadinae
Genus	—	*Pseudomonas*
Subgenus	—	(not for *Pseudomonas*)
Species	—	*Pseudomonas fluorescens*
Subspecies	—	*Pseudomonas pseudoalcaligenes* subsp. *citrulli*
Biovar	—	*Pseudomonas fluorescens* biovar I
Pathovar	—	*Pseudomonas syringae* pathovar *tabaci*

Above the rank of genus, the names should be latinized, and feminine in gender, and plural. Moreover, the hierarchy up to and including orders, is identified by means of characteristic suffixes. Thus the names of subtribe, tribe, subfamily, family, suborder and order have a stem derived from that of the type genus, and end in 'inae', 'eae', 'oideae', 'aceae', 'ineae' and 'ales', respectively (Table 6.1). These names start with a capital letter, but there is some disagreement about whether or not there is a need for italics. Essentially, italics are used in the USA, but not always in Great Britain. Although it has been emphasized that the names of higher categories are derived from the type genus, there is at least one notable exception to the rule. This example is the family Enterobacteriaceae, for which the type genus is not '*Enterobacter*' but *Escherichia*. The valid publication of a new taxon is always accompanied by deposition in a culture collection of a reference strain. This is designated as the 'nomenclatural type' or simply the 'type strain'. There is a variety of terminology for the type strain.

'Holotype' is the type strain reported by the original author. However if the original author only described one strain (this practice is not tantamount to good taxonomic procedure) and did not specifically regard it as the holotype, then it is regarded as the 'monotype'. If a subsequent investigator designates one of the original author's strains as the type culture, then it is reported as a 'lectotype'. In the situation that the original cultures have been lost, another scientist may propose a type strain, providing that it corresponds closely to the original description. This is referred to as the 'proposed neotype', and, two years after its publication in the IJSB, it is regarded as the 'established neotype'. Of course, the situation may arise that, after a neotype strain has been proposed, the original cultures are rediscovered. In this case, the Judicial Commission should be informed so that a ruling may be made. So, type strains have been delineated for each *bona fide* species and subspecies, and type species and type genera for genera and families, respectively.

The code makes allowances for the inability to change and reject names. The former could be attributed to typographical mistakes in the original description. The latter may be voiced through an Opinion of the Judicial Commission; whereupon the name is regarded as '*nomen rejiciendum*', and loses its standing in bacterial nomenclature. Conversely, a name may be conserved ('*nomen conservandum*'), in which case it must be used instead of other previous synonyms. In essence, names may be 'legitimate' or 'illegitimate' depending upon whether or not they are published in accordance with the Code. Changes may result from improvements in taxonomy. For example, a species may be transferred from one genus to another, in which case a new combination of names will result. This is referred to by the citation comb. nov. (an abbreviation for *combinatio nova*) which appears after the new name. Similar abbreviations, i.e. sp. nov. (*species nova*) and gen. nov. (*genus novum*) are used after new species or genus names. As an example, the causal agent of bacterial kidney disease in salmonid fish was elevated to new genus status, as *Renibacterium salmoninarum* gen. et sp. nov. The correct citation involves the names of the original authors and the date of the publication. This means that for the above example, the pathogen is referred to as *Renibacterium salmoninarum* Fryer and Saunders 1980.

The starting date for the priorities in bacterial nomenclature was given as January 1980. During this month, the Approved Lists of Bacterial Names were published in the IJSB. Therefore, the task of naming new taxa was made easier, insofar as reference needed to be made to this list and its subsequent supplements. Undoubtedly, the efforts described above have rationalized bacterial nomenclature, and, with universal acceptance of the Code, should radically improve the situation for bacteriology.

6.1 THE SPECIES CONCEPT

It is important to have some idea of the entities to which we are attaching names. In this context, the species has special importance because it is the only taxon which 'exists' in a population sense in the biosphere. This is particularly important with the current interest in biodiversity and the maintenance of such diversity in the environment. However, we have yet to determine how many species exist let alone attribute names to them. Moreover, the number of unculturable prokaryotes is probably very high and we may have only cultured 5% of extant bacteria (see Chapter 8).

The prokaryotic species cannot be defined easily in terms of inter-breeding as with higher organisms because of the peculiarities of prokaryotic sexual processes (but see Dykhuizen and Green (1991) who have shown how this might be achieved by DNA sequence comparisons). Consequently, alternative definitions of the species have been sought.

The traditional view of the species is a collection of phenotypically highly related strains that differ appreciably from other such groups. Gordon *et al.* (1973) recommended that such collections should include both culture collection strains and recently isolated strains so that changes associated with laboratory storage are accommodated. The numerical taxonomist is able to quantify this description by visualizing the cluster of points (strains) in a taxonomic space (see Chapter 2). The cluster can be considered as spherical with a centre (the centroid) and a radius. The larger the sample of strains (points), the more accurate the envelope of the cluster and its central point. For this reason, pheneticists suggest that species should be described optimally from collections of about 25 strains and, at the fewest, 10 strains. Such a species is best described as a taxospecies. Unfortunately, many new species have been described after the study of single isolates. This is not generally regarded as sound taxonomic practice.

A genomic species is one described by DNA sequence homology. Here the most commonly adopted standard is a collection of strains the chromosomal DNA sequence homology of which is greater than 70% measured under optimal renaturation conditions with ΔT_m <5% (see Chapter 5). Fortunately, genomic species based on this definition are almost invariably good taxospecies and the two approaches to species definition are complementary (Stackebrandt, 1992). However, problems arise in some taxa where intra-taxon DNA sequence homology is low but diagnostic phenotypic characters are not apparent. Generally, this means that the 'wrong' phenotypic characters have been assessed and further study will reveal suitable distinguishing features.

Although most strains of independent species show small differences in small subunit rRNA gene sequence these may not be suitable for species definition. *Bacillus psychrophilus* and *B. globisporus* for example, conform to the DNA homology definition of independent species but have diverged so recently that their 16S rRNA sequences are identical (Fox *et al.*, 1992). Thus rRNA sequences, so valuable at higher taxonomic ranks, may not be useful at the species level.

There is a tendency to wish to unify the species definition for all organisms and to use a concept based on inter-breeding communities for bacteria, as is done for higher organisms (Dykhuizen and Green, 1991). The genospecies (a group of organisms capable of genetic exchange) may be appropriate for bacteria if we are careful about the definition of 'inter-breeding'. Inter-breeding requires (1) the transfer of DNA between two strains followed by (2) integration of the incoming DNA into the chromosome by recombination. The latter process requires reasonably extensive sequence homology between the incoming DNA and the chromosome (and is therefore an alternative expression of DNA reassociation data). It is difficult to see how this definition of the species can be put into practice, although a population genetic approach may be the most amenable (see Chapter 5). Where species do naturally inter-breed, by for example DNA-mediated transformation as in *B. subtilis* and *B. licheniformis*, the incomimg DNA is not stably maintained in the hybrid but is excluded over a few generations and the parental species are maintained (Duncan *et al.*, 1989). It may be that insufficient homology exists between the incoming DNA and the resident chromosome for homologous recombination and the incoming DNA is lost. Inter-breeding, in this case, supports the established species limits and is an *in vivo* measure of sequence homology with the added overtones of the prevalence of natural chromosome transfer mechanisms between species, host-controlled restriction and modification of foreign DNA and recombinational proficiency of the host. This seems to apply several poorly understood criteria to our species definition and it would seem that with our current, inadequate understanding of the population genetics of bacteria that the genomic species with its robust definition based on DNA reassociation assays is the most appropriate for general usage.

REFERENCES

Duncan, K.E., Istock, C.A., Graham, J.B. and Ferguson, N. (1989). Genetic exchange between *Bacillus subtilis* and *Bacillus licheniformis*: variable hybrid stability and the nature of the bacterial species, *Evolution*, **43**, 1585–1609.

Dykhuizen, D.E. and Green, L. (1991). Recombination in *Escherichia coli* and the definition of biological species, *Journal of Bacteriology*, **173**, 7257–7268.

Fox, G.E., Wisotzkey, J.D. and Jurtshuk, Jr., P. (1992). How close is close: sequence identity may not be sufficient to guarantee species identity, *International Journal of Systematic Bacteriology*, **42**, 166–170.

Gordon, R.E., Haynes, W.C. and Pang, C.-H. (1973). *The Genus Bacillus*, United States Department of Agriculture, Washington, DC.

Lapage, S.P., Sneath, P.H.A., Lesel, E.F. *et al.* (1975). International Code of Nomenclature of Bacteria and Statutes of the International Committee on Systematic Bacteriology and Statues of the Bacteriology Section of the First International Association of Microbiological Sciences, American Society for Microbiology, Washington, DC.

Sneath, P.H.A. (1992). *International Code for Nomenclature of Bacteria (1990) Revision*, American Society for Microbiology, Washington, DC.

Stackebrandt, E. (1992). Unifying phylogeny and phenotypic diversity, in, *The Prokaryotes*, 2nd edition (A. Balows, H.G. Trüper, M. Dworkin, W. Harder and K.H. Schleifer, Eds.), pp. 19–46. Springer-Verlag, New York.

Wayne, L.G. (1991). Formal nomenclature and 'utility' terminology in bacterial systematics, in *Advances in Culture Collections* (L.H. Huang, Ed.), pp. 1–9. Pfizers Central Research, Groton, Connecticut.

CHAPTER 7

Identification and diagnosis

The identification/diagnostic processes are the end result of taxonomy. Isolates may only be identified if classification has been a preceding step and the resultant taxa given names or codes. If the organism represents an undescribed taxon, then it cannot be identified. This may appear obvious, but it is a common fault of diagnosticians who attempt to identify the unclassified. Occasionally, novel organisms are encountered that are evidently distinct from, but nevertheless resemble, recognized taxa. The favourite ploy is to label the isolate as 'presumptive' or 'atypical'. Certain outbreaks of disease described as atypical pneumonia, for example, were eventually attributed to a new taxon, *Legionella pneumophila*.

Identification is a comparative process by which unknown organisms are examined, and compared with the known (Pankhurst, 1978; Sneath, 1978). The first stage is to obtain pure cultures of bacteria. Although this need has been emphasized in earlier chapters, it is important to reiterate here the absolute requirement for pure cultures. If mixed cultures are used, the results of any identification will be quite meaningless. However, even experienced bacteriologists make the occasional mistake, especially if, for example, morphologically similar lactobacilli are being examined, all of which make small, chalky-white colonies. Cultures should be streaked and re-streaked for single colony isolation at least three times in order to ensure purity. Moreover, it is sound policy to check purity regularly by means of Gram-stained smears. These will highlight any obvious contamination. Thereafter, the pure culture is ready for the onslaught of the diagnostician.

Cowan discussed three views for identifying medical bacteria (see Barrow and Feltham, 1993). As there is great truth in those statements, it is appropriate to consider them here, At one extreme (the 'blunderbuss approach'), the unknown isolate can be examined for all possible tests. When the results are available, use is then made of standard texts, such as *Bergey's Manual of Systematic Bacteriology* (Krieg, 1984). Although this approach is time consuming and may be expensive in materials, it is envisaged that the battery of test results will permit the ready identification of the organism. In practice, some important diagnostic tests are usually forgotten, which leads to the necessity for more work and further time is wasted. This approach to identification is widely used, but it is not suited for the very busy diagnostic laboratory, where time and personnel are at a premium. Conversely, there is a step-wise approach in which diagnostic schemes are followed progressively as used in *Cowan and Steel's Manual for the Identification of Medical Bacteria* (Barrow and Feltham, 1993). Thus answers for the first tests will allow the diagnostician to proceed with the next series of tests, until an answer is obtained. Again, this may be highly time-consuming. Thirdly, there will be many occasions during which the diagnostician has reasonable grounds to suspect the identity of an unknown organism. Therefore, specialist schemes suitable for identifying the category of organism suspected may be consulted. Such occasions might include recovery of cultures on selective media, or from well-described pathological conditions. In these instances, use is made of the experiences, and perhaps intuition, of the diagnostician. Mistakes will occur, but rarely will the judgement be questioned.

So far, oblique reference has been made to the use of tests involved in identification processes. Many of these are classical bacteriological tests, such as the Voges–Proskauer reaction, whereas others represent more recent developments in microbiology, such as determining the presence of specific subcellular components. The results for such tests are compared with data for known organisms. The absolute requirement for carefully executed testing regimes should therefore be apparent; reliable and reproducible methods are necessary. The discussion of test error and reproducibility in Chapter 2 applies equally well here, insofar as error will inevitably lead to erroneous identification. However, the error will be considerably reduced if the tests are carried out properly. Any short cuts or bad techniques will reduce the effectiveness of the identification process. Attention to detail is a prerequisite of sound identification practice. This is an opportune moment to consider the nature of the tests to be used. These will hopefully be sensitive tests giving clearly defined reactions depending upon the ingenuity of the scientist who initially constructed the scheme. Considering that a reaction is dependent upon the nature of the methods used, it is important to mimic, as closely as possible, the precise methods used in the construction of the scheme. For example, if motility was initially determined from wet preparations after 18 h incubation in nutrient broth, the diagnostician should use the same method. In this case, motility at 18 h could appear to be negative after 24 h. A wrong result would be recorded, which could contribute to misidentification. This has importance for the differentiation of closely related motile and non-motile species, such as *Aeromonas hydrophila* and *A. salmonicida*.

7.1   SEQUENTIAL IDENTIFICATION SYSTEMS: DICHOTOMOUS KEYS

Dichotomous keys (diagnostic keys) were among the first forms of bacterial identification schemes. Indeed, they may be found in the earlier editions of *Bergey's Manual of Determinative Bacteriology*. These are sequential systems in which identification is achieved in a step-wise fashion. The diagnostician proceeds along progressively branching routes depending upon positive or negative reactions to tests. If the test is positive then one route is followed but if a negative result is recorded then an alternative pathway is used (see Fig. 7.1) There are two main drawbacks to this process. The first is that microbiological

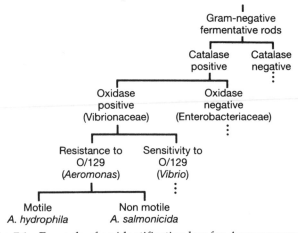

Fig. 7.1   Example of an identification key for *Aeromonas* species.

characters are not normally gathered in a sequential fashion. With flowering plants, for example, the information for identification can largely be gathered by visual inspection. Thus, the questions can be answered in one sitting, and in a sequential fashion by examining the specimen on the laboratory bench. Bacteriological data, on the other hand, result largely from growth tests that may take anything from one day to two weeks or more to complete. It therefore becomes imprac-tical to be answering such questions in a stepwise manner and to require the result of one test before the next can be started.

The second drawback of sequential keys arises from an incorrect result or an organism with an aberrant feature, both of which will send the diagnostician along the wrong branch of the key, leading to mis-diagnosis. Indeed, if the probability of a correct test result is 90%, in a key comprising six tests there will be a probability of about 50% that the wrong answer will be given by the time the tests have been completed. Largely for this reason, dichotomous keys have lost favour to diagnostic tables, and only a few keys have been proposed since

Table 7.1 Diagnostic table for identifying *Aeromonas*

Character	Aeromonas hydrophila	Aeromonas media	Aeromonas salmonicida
Gram-negative fermentative rods	+	+	+
Presence of brown diffusible pigment	–	+	+
Growth at 37°C	+	+	–
Indole production	+	V	–
Degradation of casein	+	V	–
Motility	+	–	–
Phosphatase production	+	–	V

+, – and V correspond to ≥ 80%, ≤ 20%, and 21–79% positive responses, respectively.

1970. Therefore, it must be concluded that the older keys lack the benefits of more recent developments in bacterial taxonomy.

7.2 SIMULTANEOUS IDENTIFICATION SYSTEMS: DIAGNOSTIC TABLES

Diagnostic or identification tables employ a simultaneous approach to identification. An end result of numerical taxonomy studies should be the construction of diagnostic tables (e.g. Table 7.1). These lack the disadvantages of dichotomous keys because they are polythetic (use overall matching of the unknown with the reference rather than presence or absence of key features) and the presence of a few aberrant results is less likely to influence the outcome of the identification process adversely. With diagnostic tables, it is recognized that not all members of a taxon will give uniform test results. Thus, the presence of + or – in the table indicates that most isolates correspond to these results. Usually + and – correspond to 80–85% and 15–20% positive responses, respectively. Results indicated as V or D indicate variable positive responses, i.e. in the range 21–79%. Diagnostic tables, which contain a matrix of test results for a range of bacterial taxa, may be found in widely used texts, such as *Cowan and Steel's Manual for the Identification of Medical Bacteria* (Barrow and Feltham, 1993) and *Bergey's Manual of Systematic Bacteriology* (Krieg, 1984). The sequential approach to identification offered by tables is eminently suitable for microbiological purposes. The unknown isolate is examined for the tests indicated in the diagnostic table. These can be generated as a battery of growth or biochemical tests. Then, the results are compared with the profile for each taxon. Obviously for large comprehensive tables, this may be a time-consuming process and it is often difficult to obtain accurate diagnosis by eye. However, there is scope for semi- and fully-automated procedures, such as offered by punch cards and computers respectively.

7.3   COMPUTER-BASED IDENTIFICATION SYSTEMS

It is possible to use diagnostic tables in conjunction with computers. In the simplest form, the tables are incorporated into the computer memory, and the results for unknown organisms are compared with each of the possibilities and the highest correlation is taken as an identification.

Payne (1963) has been credited with the initial development of computer-based identification procedures. Thereafter, the work of Dybowski and Franklin (1968) is noteworthy, insofar as these workers successfully identified medically important Gram-negative bacteria by use of computer techniques. In fact, a 50% success rate (6 of 12 strains) was recorded. From this modest start, Lapage and co-workers from the Central Public Health Laboratory (CPHL), Colindale, UK, successfully achieved reliable computer-based identification systems for 'hard-to-identify' Gram-negative bacteria of medical importance. Lapage *et al.* (1973) described a computer-based identification matrix comprising 62 reference taxa and their responses to 50 phenotypic tests commonly used in medical diagnostic laboratories. These included fermentation of 18 sugars and other characters known to be useful for fermentative Gram-negative bacteria, such as decarboxylase reactions, H_2S production, the methyl red and Voges–Proskauer tests.

A data matrix was constructed in which each test result was converted into an estimate of the probability of a positive response. Thus, values of 0.99 and 0.01 were allocated for results that were always positive and negative, respectively. Results that were usually positive or negative were scored as 0.95 and 0.05, respectively. Values between 0.05 and 0.95 were attributed to variable test results depending on the frequency of positive results expected for the taxon. In these cases, the values were deduced after consulting published literature and Public Health Laboratory Service records. A computer program was devised based on Baye's theorem, which sought to identify isolates in three stages. These included the calculation of the likelihood that the unknown belonged to a particular taxon, determination as to whether or not a definitive identification had been achieved and the selection of any further tests necessary to improve the chance of successful identification.

The likelihood of an unknown isolate belonging in a given taxon is defined as 'the probability of obtaining the observed test results with a strain of this taxon'. This probability is derived by multiplying together the probabilities of the individual test results (see Fig. 7.2). Thus, the computer compares the results of the unknown organisms against the entry for each taxon contained in the matrix. Where the unknown organism has a positive test result, the probability value is taken directly from the matrix. For example, if the unknown isolate produces catalase then during the comparison with a given taxon the probability of a positive result with this test is included in the calculation. However in

Data matrix:

Taxon	Probability of positive result for:		
	Catalase production	Oxidase production	β-Galactosidase production
1	0·80	0·95	0·05
2	0·99	0·01	0·95
Unknown isolate 'x'	+	−	+

Calculations:

Probability products("likelihoods")

'x' compared with taxon 1: $0·80 \times (1-0·95) \times 0·05 = 0·002000$

'x' compared with taxon 2: $0·99 \times (1-0·01) \times 0·95 = \underline{0·931095}$

Sum = 0·933095

Percent relative likelihoods

'x' compared with taxon 1: $0·002000/0·931095 = 0·0021480085$

'x' compared with taxon 2: $0·931095/0·931095 = 100\%$

Identification score (normalized)

'x' compared with taxon 1: $0·002000/0·933095 = 0·0021434$

'x' compared with taxon 2: $0·931095/0·933095 = 0·9978566$

Fig. 7.2 Determination of the identification score for bacteria (after Lapage *et al.*, 1973). See text for details.

the case of negative results for an unknown isolate, the probability is obtained by subtracting the probability of a positive response from unity (Fig. 7.2). The 'likelihood' figure gives an indication of which taxon the unknown isolate most closely resembles, but this may be more than one taxon and, therefore, a definitive identification is not obtained. Dybowski and Franklin (1968) approached this problem of ranking likelihoods by calculating the percent relative likelihood in which the relative likelihood for each taxon is divided by the maximum likelihood obtained (Fig. 7.2). This is useful but still does not provide a critical level above which unequivocal identification can be assumed because the unknown may match two entries in the matrix very closely and then the choice between them is problematic. Lapage *et al.* (1973) 'normalized' the identification score to provide an unequivocal identification. By dividing the likelihood for each taxon by the sum of likelihoods for the matrix (normalizing), a figure is obtained that approaches 1.0 if the unknown identifies closely with one taxon and very little with any others. If it matches two entries in the matrix the normalized score will be low and identification is not achieved. This normalized score is often referred to as a 'Willcox probability' after Dr.

Willcox, who first applied the procedure. By choosing an identification score of 0.999, it can be safely assumed that any isolate achieving this value must resemble only that one taxon and no others. Problems can arise, however, if the taxon to which the unknown belongs is not contained in the matrix. The unknown may then be identified with high normalized probability to an unrelated taxon. This possibility must always be guarded against. This can be done by using one's own judgement; very often such grossly incorrect identifications can be detected by visual examination of the bacterium. Alternatively, some other statistical measure of identification that does not rely on estimates of likelihoods, for example estimation of taxonomic distance or standard error of taxonomic distance can be used. These are discussed in detail by Willcox *et al.* (1980) and Priest and Williams (1993); suffice it to say that taxonomic distance can be interpreted as the negative logarithm of the likelihood and is a useful complement to the normalized likelihood, because in instances where the latter is erroneously large due to the absence of the taxon from the matrix, taxonomic distance will be unacceptably large and indicate that the Willcox probability is performing inadequately. Standard error of taxonomic distance attempts to set an envelope to the taxon and estimates the distance of the unknown from the centroid relative to the mean distance. A low value indicates identification.

The level of Willcox probability required for a 'correct' identification can be set according to the organisms under study. Lapage *et al.* (1973) were able to use 0.999 with fermentative Gram-negative bacteria, because these are well-studied organisms with tight homogeneous species. With less well-described taxa, such as streptomycetes where overlap between taxa is more prevalent and species are less well defined, an identification score of 0.85 is more realistic (Williams *et al.*, 1983). For *Bacillus*, we routinely use 0.95 but supplement the Willcox probability with a measure of taxonomic distance to substantiate the identification.

Probabilistic identification by computer can also be treated in a sequential fashion. Not all the tests in the matrix are required to be completed for computation of identification scores. If only seven of ten tests have been completed, a trial identification can be attempted. If an acceptable Willcox probability is achieved, some work has been saved. If the probability is too low, however, the computer can be programmed to compute the diagnostic value of the remaining tests (essentially by calculating their discriminatory power for distinguishing the few most likely taxa) and listing these in order of discriminatory value. Some or all of these tests can be conducted followed by a second attempt at identification. Thereafter, a third series of tests and subsequent identification could be carried out if necessary. This ranking of tests according to diagnostic value can be extremely helpful and greatly reduces the amount of work necessary to effect an identification.

The computerized identification system, derived by Lapage *et al.* (1973), was a success insofar as 81.4% of problematical Gram-negative organisms were identified correctly. When a limited number (i.e. 30) of tests was used, the success rate was 77.4%. Some problems were uncovered, but these were resolved by modifying and extending the data matrix. Thus, in a series of articles published in 1973, the matrix contained data for 51 tests on 70 reference taxa (Bascomb *et al.*, 1973; Lapage *et al.*, 1973). In these studies, 1079 reference cultures and 516 problematical fresh isolates of aerobic Gram-negative rods were examined against the data matrix. The outcome was that for fermentative organisms, 90.8% of the reference strains and 89.4% of the fresh isolates were identified. Less success was recorded with non-fermenting organisms, insofar as identification was achieved with only 82.1% of the reference strains and 70.8% of the fresh isolates. However, new matrices specifically designed for these organisms have since been prepared. Obviously, the future is bright for computer identification systems, and today systems based on the CPHL approach are used worldwide, particularly for identification of Gram-negative and Gram-positive bacteria of clinical importance. Moreover, there was considerable potential for commercial exploitation and most producers of identification kits offer identification services based on the calculation of Willcox probabilities.

7.4 SEROLOGY

Much confusion has resulted from the use of serology for bacterial identification, as may be deduced from the several hundreds of *Salmonella* 'species' included in the seventh edition of *Bergey's Manual of Determinative Bacteriology*. Yet treated cautiously, a wealth of information may be derived from the reaction of antigen with antibody. Numerous serological reactions have been described, including whole-cell agglutination (WCA), latex agglutination, precipitin reactions, direct and indirect fluorescent antibody tests, immunohistochemistry and the enzyme-linked immunosorbent assay (ELISA) (see Smibert and Krieg, 1981). The main advantage is speed, and in some cases, such as with WCA, diagnosis may be achieved within a few minutes. With certain methods, e.g. ELISA, there is the potential for the development of kits suitable for field use, particularly directly on diseased tissue and without the need for isolating the offending organisms. In short, serology may be used to identify bacterial cultures, bacterial subcellular components, or bacteria embedded within tissues. The requirement is for reliable antibody (contained in antiserum).

In their crudest form, antibodies are raised in mammals, such as rabbits, rats, guinea pigs or mice, following injections with the antigen, i.e. whole bacterial cells or purified subcellular components. With time, the mammalian antibody-producing cells (β-lymphocytes) secrete antibody into the blood, which may be removed, allowed to clot and thus give rise to antiserum. In the presence of sodium chloride, antisera react with antigens to give rise to a measurable reaction. If the reaction is homologous, i.e. the antibody associates with the antigen against which it was prepared, then the response is useful for diagnosis. However with heterologous reactions, the antibody cross-reacts with different antigens, and mis-diagnosis may ensue. The problem is associated with the complex antigenicity of the bacterial cell, and the presence of non-specific shared antigens. To some extent, heterologous reactions may be reduced by using antisera prepared against specific antigens, which are known to be restricted to the bacterial taxon under study. This requires a detailed knowledge of bacterial biochemistry, which is lacking for most taxa. Specificity of antisera may also be improved by development of monoclonal antibodies, which are now being used more extensively. For the present, it is conceded that serology provides useful diagnostic information, which should not be used in isolation. Serological diagnosis of bacteria should be confirmed by other phenotypic traits. Nevertheless, serology is of paramount importance in epidemiological studies for determining the serotype (or serovar) of any organism.

7.5 COMMERCIAL KITS

A cursory market survey would reveal the considerable potential for sales of identification kits, especially those aimed at medical laboratories. Indeed, many manufacturers have developed kits of which a few types have gained widespread use. However, most of these products have been designed for specific purposes, principally identifying certain groups of medically important bacteria. Within these constraints, it would appear that commendable success has been achieved, but problems may ensue when the kits are used for other purposes, such as for identifying a wide range of environmental isolates. In this context, it is important to note that the natural environment contains a wider range of taxa than that encountered in medical diagnostic laboratories.

Essentially two types of commercial kits have been developed; those that focus on biochemical responses of the organism, and those based on serological reactions. To date, the former have taken the larger share of an expanding market. For the future, there will be the exploitation of gene probe technology, which is currently being developed at a rapid rate (see section 7.8).

Commercially available kits measuring biochemical activity centre around rows of microtubes or paper strips impregnated with various freeze-dried test substrates. These are rehydrated by inoculation with bacterial suspensions, and, after a pre-determined incubation period, the results are recorded as colour changes usually following addition of reagents. Identification may be achieved within 2–48 h. The simplest forms of kit, measuring a maximum of three biochemical tests, were marketed under the trade names of Dip slide (Oxoid), Microstix (Miles Laboratory) and Minitek (Becton Dickinson). These kits were designed for Enterobacteriaceae, *Neisseria gonorrhoeae* and human pathogenic *Neisseria*, respectively. More comprehensive kits, which measure as many as 50 biochemical reactions, have been developed, and include the API systems (API Laboratory Products), Enterotube and the Oxi-Ferm tube (Roche), Minitek (Becton Dickinson) and Pathotec or the Micro-1D system (Warner Lambert). Variants of these products have been developed for identifying anaerobes, *Bacillus*, Enterobacteriaceae, lactobacilli, pseudomonads, *Staphylococcus*, *Streptococcus* and yeasts. With most of these systems, use may also be made of a computer-based data matrix for identifying unusual or atypical isolates. The oldest and one of the most widely used products is the AP1-20E rapid identification system, which was first marketed in 1969. This kit comprises 20 biochemical reactions, namely β-galactosidase, arginine dihydrolase, lysine and ornithine decarboxylases, utilization of Simmons citrate, H_2S production, urease, tryptophan deaminase, indole production, the Voges–Proskauer reaction, gelatinase, and acid production from glucose, mannitol, inositol, sorbitol, rhamnose, sucrose, melibose, amygdalin and arabinose. In addition, the oxidase test needs to be recorded. The microtubes are re-hydrated with a few drops of the bacterial suspension, sterile mineral oil is added to cover some tubes, and the tray incubated at 35–37°C for 24 or 48 h. The reactions are then recorded, in some cases after the addition of reagents. Thus, there will be results for a total of 21 tests (including the oxidase test). These are coded numerically using an octal code system into a unique 7-digit number for comparison with an index. The steps have been outlined in Fig. 7.3. It is our experience that the API-20E rapid identification system compares very favourably to the other commercially available products. A more rapid version based on enzyme reactions, which permits identification within 4 h, has been developed.

Enterotubes differ from API systems insofar as the former are inoculated by means of a wire that is inoculated with culture and then drawn lengthways through the compartments containing the already hydrated media. Minitek comprises wells into which substrate containing discs are added. In contrast, Pathotec uses paper strips, which are impregnated with substrates. These are added to a bacterial suspension with incubation for 4–6 h. Biolog uses dried substrates in 96-well microtitre trays which contain an indicator of carbon source metabolism, a triphenyltetrazolium salt. These therefore indicate carbon

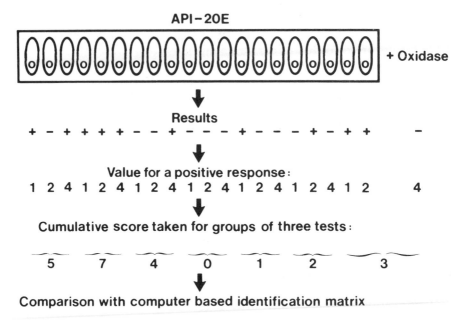

Fig. 7.3 Use of API-20E system for the rapid identification of bacteria.

source utilization rather than acid production in a fermentation reaction. Some systems, e.g. API-zym and the Sensititre scheme are based entirely on the detection of bacterial enzymes such as the ONPG reaction for β-galactosidase, a task which may be accomplished within 24 h and with automation.

The speed and ease with which these kits may be used is an obvious advantage. Thus the savings in media preparation and space compensate for the seemingly high cost of the kits. Nevertheless, there are problems with false-positive or false-negative activities. Of course, this criticism could also be made against the comparative conventional tests and in many cases the kits are more reliable. One obvious development has been the manufacture of identification machines. One particular product is based on fluorimetric enzyme substrates dispensed as solids in the wells of sterile 96-well microtitre trays. Addition of culture to these wells resuspends the substrates and the tray is incubated in a dedicated incubator. After a short period, the fluorescence in the wells is read using an appropriate detector and the results recorded as positive or negative against a particular threshold value. The pattern of test results is then matched with a database using probabilistic identification statistics. A second machine that is currently available uses similar technology but colorimetric substrates based on triphenyltetrazolium salts held in a microtitre tray. Again patterns of results are recorded automatically and compared with suitable databases held in the instrument. These systems work well for Gram-negative bacteria of clinical significance (for which they were developed) but are less suitable for identification of Gram-positive bacteria such as streptomycetes or bacilli.

Serological tests are gaining popularity, particularly because of their ease of use and sensitivity. Kits have been developed for detecting a wide range of organisms, including *Bacteroides* spp., *Neisseria* spp., *Haemophilus influenzae,* and *Streptococcus* spp., by latex agglutination and immunofluorescence techniques. Moreover current research is aimed at developing and evaluating monoclonal antibodies destined for incorporation into the ultra-sensitive ELISA. Therefore, it is anticipated that the range of serological kits will continue to be popular.

7.6 BACTERIOPHAGE TYPING

The primary interest with bacteriophages is in epidemiological investigations. Thus the bacterial viruses are of great value in determining the presence of specific sub-groups (types) of organisms. Indeed, bacteriophage typing schemes have been proven to be useful for many bacterial pathogens, including *Aeromonas salmonicida* and *Staphylococcus aureus.* The technique involves use of virulent (lytic) bacteriophages, which are inoculated by dropping onto freshly seeded bacterial lawns. After incubation for 18–48 h, a positive result is recorded by the presence of zones of clearing (plaques) in the otherwise uniform layer of bacterial growth. The reactions between bacteriophage and bacteria appear to be extremely specific. In practice, bacteriophages usually lyse strains from the same taxospecies, occasionally from other species from within the same genus, and extremely rarely from strains of different genera. However, it is unclear how much emphasis should be placed on the few apparent cross-reactions.

7.7 CHEMOSYSTEMATIC METHODS IN IDENTIFICATION

Many chemosystematic methods are eminently suitable for identification purposes, particularly those that provide quantitative data which can be analysed by computer. Thus electrophoretic protein profiles, fatty acid profiles and Py-MS (see Chapter 4) are being developed for identification and have the advantages of speed and ease of automation over conventional tests.

Essentially, the approach involves the determination of profiles for reference strains, and either using these to prepare a classification or using an existing classification and storing the resultant data matrix in a computer. Profiles for unknown organisms may then be compared with patterns for reference taxa and identification is achieved if the patterns correlate. Of the three methods, electrophoretic protein profiles have been successfully applied to various groups of clinical and ecologically important microorganisms. The method requires reasonably sophisticated electrophoresis apparatus and to obtain the necessary reproducibility, preformed polyacrylamide gels are best. The method then provides accurate distinction of groups with the resolving power of DNA hybridization. Thus, it is best suited to the delineation of the species, subspecies and biotype within bacterial genera (see Fig. 4.1 in Chapter 4).

Fatty acid analysis is also gaining popularity in bacterial identification because of the ease with which it can be automated. In particular, an identification machine using gas chromatography of fatty acid methyl esters is available from Hewlett Packard. The production of a profile takes about 60–90 minutes (60 minute sample preparation time) but when samples are prepared in batches the processing time is much lower and quoted at 4 minutes per sample. The machine compares the trace with a library of traces stored in a computer to effect an identification.

One method which offers the advantages of speed and automation is pyrolysis-MS (see Chapter 4). Mass spectrometry of pyrolysed microorganisms results in complicated profiles of mass ions which differ between strains and species only in rather minor ways (see Fig 7.4). Therefore data analysis to provide discrimination between strains of different species is complex. The data analysis is based on various discrimination procedures in which a combination of the quantitative

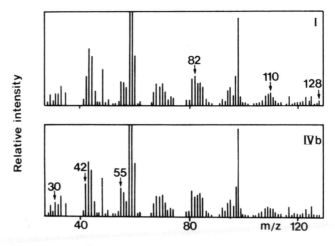

Fig. 7.4 Mass spectra of two *Listeria* serotypes (I and IVb). Masses responsible for discrimination are arrowed (from Gutteridge, 1987, with permission).

characters that best differentiates an established set of taxa is determined. It is important to note that the taxa are not produced by these procedures (i.e. discriminant analysis does not classify the organisms); the groups must first be determined by hierarchical clustering or some ordination procedure. Discriminant analysis simply defines those characters that best distinguish the taxa.

One type of discriminant analysis is canonical variates analysis, which is similar to principal components analysis (see Chapter 2). A transformed axis is determined which seeks to maximize the ratio of the variability between the means of the different taxa to that within the taxa. A second canonical variate is then sought orthogonal to the first and representing the next greatest variability, and subsequent axes are then derived. From these, a set of canonical variate means for each taxon can be determined and, for an unknown, a set of canonical variate scores is produced. For identification, the Euclidean distance of the unknown from each of the taxa can be determined and the organism assigned to the closest (MacFie *et al.*, 1978).

Stepwise discriminant analysis may be used to determine those characters that maximize the ratio of variance between, to that within, groups. A subset of discriminant functions is calculated that provides a stable solution and these classification functions can be used to identify new samples. In the example shown in Fig. 7.5 and Table 7.2, canonical variates analysis of Py-MS data shows that strains within the taxa *Bacillus amyloliquefaciens*, *B. pumilus*, *B. lichenformis* and *B. subtilis* could be discriminated according to this classification. Using stepwise discriminant analysis, 16 MS masses were needed to distinguish these groups effectively. Eight unknown strains were then analysed and correctly identified (Shute *et al.*, 1985). One problem with Py-MS is drift in the machine. Over prolonged periods (several months) the sensitivity of the machine changes and repeats of standard materials do not give identical results. This causes problems when identifying

Table 7.2 Identification of eight unknowns using stepwise discriminant analysis (taken from Shute *et al.*, 1985 with permission)

	Distance from group mean				
Actual identity	*S*	*P*	*L*	*A*	*Identification*
B. subtilis (S)	3.7	19.6	17.2	22.2	S
B. subtilis (S)	14.5	27.8	41.4	15.9	S
B. pumilus (P)	29.5	6.9	26.6	30.8	P
B. pumilus (P)	45.5	13.0	43.1	33.4	P
B. licheniformis (L)	24.6	24.4	7.2	38.4	L
B. licheniformis (L)	10.0	16.1	7.5	27.1	L
B. amyloliquefaciens (A)	39.8	20.8	32.5	11.6	A
B. amyloliquefaciens (A)	87.6	58.2	91.6	32.5	A

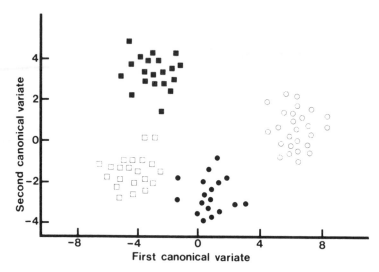

Fig. 7.5   Canonical variates analysis of four groups of *Bacillus* strains: ●, *B. subtilis*; ○, '*B. amyloliquefaciens*'; ■, *B. pumilus*; □, *B. licheniformis* (from Shute *et al.*, 1985, with permission).

isolates against libraries of stored data, since even the reference strains will not identify. This drift problem requires that unknowns are always examined with standards so that comparisons can be made at the same time. Attempts to overcome this problem by calibration of the machine with internal standards are in progress.

The possibilities for automated rapid identification of micro-organisms based on chemosystematics currently focus on Py-MS and fatty acid analyses. Both provide identifications that are as good as the databases on which they are based. The former is better suited for typing individual isolates, in epidemiology studies for example, and the latter for strain identification at the species level. With improvements in technology and the statistical handling of data, the possibilities for automated rapid identification are bright.

7.8   HYBRIDIZATION PROBES

Applications of molecular biology, particularly nucleic acid hybridization techniques, have revolutionized microbial identification through the development of nucleic acid probes. If two organisms differ pheno-typically, this is a reflection of differences in genome sequence and this difference can be detected, at least theoretically, by a hybridization probe. At first, the possibilities for this technology seemed boundless, and many companies and laboratories enthused about the coming diagnostic revolution. There followed a period in the late 1980s of disillusionment when specificity and limits of detection seemed not to live up to expectation, but with improved technology leading to greater sensitivity and higher specificity, probes are again proving to be very popular.

Hybridization probes have several applications. They can be used to type organisms by, for example, RFLP analysis of rRNA genes (see Chapter 4). When side by side comparisons are done, or patterns are recorded and compared with a library of such patterns, this approach can be used for identification. The patterns are generally simple and not suitable for identification of an unknown against a wide range of genera. This is more of a typing scheme where the approximate identity of the unknown is recognized and allocation to subspecies or biotype is required.

The major applications of probes are in the simultaneous detection and identification of micro-organisms. These two processes are intimately entwined in probe technology since successful detection of a microbe in a sample using hybridization means that complementary sequences are present and by inference some indication of the identification of the gene or organism is provided. The specificity, however, is sometimes compromised in detection probes in order to maximize sensitivity. On the other hand, a probe which provides successful identification and has high selectivity may not be very good for detection purposes because of poor sensitivity. Nevertheless, for simplicity we shall consider both detection and identification together.

The hybridizations are normally done as immobilized assays on nitrocellulose or nylon membranes. The reactions are usually semi-quantitative (i.e. strong, weak or no hybridization) but can be made quantitative by careful attention to procedures, and use of a detection system that allows quantitation by densitometric scanning or radioactive counting. To assist in quantitation and to permit large numbers of samples to be processed, the target DNA, which is usually chromosomal DNA, is usually immobilized using a vacuum filtration manifold which allows the accurate allotment of DNA samples (up to 96 depending on the apparatus) to a membrane in a set pattern. Dot-blots use an array of circular spots on the membrane. Slot-blots use an array of oval shaped spots (Fig. 3.4) which are considered to be better for subsequent quantitation by densitometric scanning.

7.8.1 The probe

The probe can be directed at a variety of sequences depending on the purposes of the test (Table 7.3). There are two broad classes of probe, those developed empirically and those directed at a specific target. The simplest of empirical probes uses whole cell DNA. As discussed in Chapter 3, strains within a species should show at least 70% DNA

Table 7.3   Some probes for detection and identification of bacteria

Probe	Target	Bacterium
Empirically designed probes		
1. Whole cell DNA	Chromosomal DNA	*Bacteroides* strains *Campylobacter* strains
2. Cloned genomic fragments	Chromosomal DNA	*Bacteroides fragilis* *Campylobacter jejuni* *Mycobacterium leprae* *Mycobacterium tuberculosis* *Mycoplasma* species
Directed probes to:		
1. Surface epitopes	Genes for surface antigens	*Campylobacter jejuni*
2. Virulence factors		
Cloned toxin genes	Mosquito larval toxin genes	*Bacillus sphaericus*
Cloned toxin genes	Enteric toxin gene	*Escherichia coli*
Synthetic oligonucleotide	β-Haemolysin gene	*Listeria monocytogenes*
Synthetic oligonucleotide	Enterotoxin B gene	*Staphylococcus aureus*
3. Conserved genes[1]		
Synthetic oligonucleotide	Variable regions of 16S rRNA	*Neisseria* species
Synthetic oligonucleotide	Variable region of 23S rRNA	*Mycoplasma* species

[1] A full list of more than 50 applications of rRNA probes for identification of prokaryotic and eukaryotic micro-organisms is given by Ward *et al.* (1993).

homology and strains from a separate species will generally have less than 50% sequence homology with other species. This then forms the basis of a reasonably specific probe. By labelling whole cell DNA and hybridizing it to immobilized whole cell DNA from reference and unknown strains, the differences in sequence homology can be exploited and reasonably good identification of strains achieved (see Fig. 3.4). Greater than 70% sequence homology will provide strong hybridization signals for members of the same species but problems arise from closely related species that pose 50–60% homology. Such strains would be readily detected by this procedure and the relatively strong hybridization reactions might be misinterpreted as identity with the probe DNA.

A refinement of the whole cell DNA approach, which avoids the problems of strong reactions from closely related species, is to clone random fragments of DNA from the reference strain and to test these for hybridization to the reference strain and lack of hybridization to other related species. In this way, fragments of DNA can be obtained that have proven specificity for the reference strain and known discriminatory power. Critics of this approach dislike the 'unknown' properties of the probe and prefer to work with characterized DNA fragments.

The large number of cloned genes and gene sequences available today can be exploited for probe design without recourse to the empirical approaches. Such probes can be targeted to certain genes in the chromosome or plasmids to provide detection and identification of specific properties. Not only can *E. coli* or *Staphylococcus aureus* be detected, but the system will also indicate the potential toxicity of the identified strain thus saving time and effort in determining the relevance of a particular strain to a clinical subject. The first, and often most successful, targeted gene probes used cloned toxin genes. Thus, DNA could be extracted direct from a faecal sample or from a colony on a plate and hybridized to the cloned enterotoxin gene DNA from *E. coli*. A positive reaction identified the bacterium as a toxigenic *E. coli* or closely related intestinal pathogen, such as *Shigella*. The number of applications of this approach is now large, both in clinical and environmental microbiology. We use it to detect strains of *Bacillus sphaericus* that are toxic towards mosquito larvae and are employed as biocontrol agents to reduce larval populations in mosquito breeding sites. The cloned toxin genes hybridize specifically to chromosomal DNA from *B. sphaericus* mosquito pathogens allowing us to detect and identify colonies from isolation plates by hybridizing total DNA to the probe in dot or slot-blots (see Fig. 3.10). Some other possibilities are listed in Table 7.3 and are reviewed by Sayler and Layton (1990).

Where the cloned gene is not available but the gene sequence has been published, it is possible to make a synthetic oligonucleotide which will hybridize specifically with the gene of interest. Such oligonucleotides should be at least 15 bases long and preferably nearer 25 or 30 so that non-homologous regions do not destabilize hybrid formation. Fifteen bases would be fine if there was no sequence divergence between probe and target but this is an unlikely situation when detecting genes in natural populations. Areas of high sequence conservation can be targeted when divergent genes are being detected. Oligonucleotide probes are more difficult to handle than larger, cloned gene probes; the shorter length requires close attention to hybridization conditions to obtain a stable duplex structure and they are more difficult to label to high specific activity than longer stretches of DNA (see Stahl and Amann, 1991 for methodological details). However, improved labelling techniques in the form of kits of reagents for end-labelling oligonucleotides are making them easier for the non-specialist to handle.

For many purposes, the cloned gene approach is too specific. The cloned enterotoxin gene from *E. coli* for example would detect and identify toxigenic strains but would not react with non-toxic varieties.

In this case, a species-specific probe is desirable. Such probes are usually directed at the rRNA molecules or their genes. These genes have several advantages as targets. The rRNA genes contain both highly conserved sequences and more variable regions (see Chapter 3). Probes are generally synthetic oligonucleotides directed towards certain sequences. For species identification, it can be directed at the variable regions within which there is normally sufficient divergence between strains of different species to be able to find a sequence specific for a species or even a subspecies. Test kits for mycoplasmas, *Legionella* species and mycobacteria have been available for several years based on this approach and with the ever-increasing number of rRNA sequences available it should continue to be refined and expanded into other groups. There are some situations where strains of different species have diverged so recently that the 16S rRNA genes are identical. This is the case in some psychrophilic bacilli and in the *Bacillus anthracis, B. cereus, B. thuringiensis* group, so this approach may not be universal.

Just as the variable regions of the 16S rRNA genes can be used for species identification, so more highly conserved regions may be used for identification of strains to higher groups. Indeed, the full range of hierarchical conserved sequences allows probe design up to, and including, the differentiation of domains and kingdoms. So archaea and bacteria can be distinguished by the appropriate sequences in 16S rRNA genes.

The second advantage of rRNA as targets for identification probes is the large copy number of these molecules. Each bacterium contains several thousand ribosomes and each is a target for the probe. The sensitivity of the system is therefore much higher than for chromosomal sequences. This is of lesser importance since the introduction of the PCR for amplifying DNA.

7.8.2 The target

Having prepared a probe, let us now examine the opportunities for its applications. One of the most frequently used target nucleic acids is in the form of colony hybridizations. Colonies on a plate are transferred to a membrane and then lysed and processed to remove protein and leave the denatured nucleic acids as single strands bound to the support ready for hybridization to the probe. A common problem with this procedure is high backgrounds due to contaminating materials, but these can normally be minimized by modification of the processing conditions. This method has been used for numerous environmental studies, such as the detection and enumeration of bacteria in environmental samples ranging from soils to rumen contents from animals. It also has application in the detection of plasmid maintenance in released micro-organisms and in particular genetically engineered organisms

(GEMS). Colony hybridization to appropriate probes has been used to monitor mercury resistant bacteria in contaminated environments, naphthalene-degrading bacteria in hydrocarbon-contaminated soils and enterotoxic *S. aureus* in foods and for many other applications (see Sayler and Layton, 1990).

If colony hybridizations prove to give high backgrounds and loss of resolution, it is advisable to prepare DNA or total nucleic acids from colonies and to blot this material onto the membrane using a filtration manifold. Total nucleic acids can be prepared on a small scale with all manipulations performed in microfuge tubes since only small amounts of DNA (a few micrograms) are needed. In this way, large numbers of samples can be prepared and blotted in a matter of hours.

One of the major breakthroughs in environmental microbiology has been the application of probe technology for the detection and identification of micro-organisms directly in the environment (see Chapter 8). Total nucleic acids are extracted from soil or seawater and hybridized to appropriate probes. Positive signals indicate the presence of micro-organisms in the sample, which can be identified according to reactions to specific probes. In this way so-called 'unculturable' micro-organisms can be detected and identified at least to genus or higher rank and the technique is providing a new insight into the diversity of microbial populations. Extraction of the nucleic acids from the sample is the problem. Water samples are relatively straightforward, large volumes of water are filtered and the collected cells lysed and the nucleic acids prepared in the normal way. Cell lysis is often a problem with Gram-positive bacteria and when the sample contains unknown microbes this difficulty is more severe. Mechanical breakage of cells circumvents this drawback. Extracting nucleic acids from soils is more difficult because clay particles and humic acids remain tightly bound to DNA as impurities. Two approaches are commonly used. Cells are removed from the soil by exhaustive washes and extraction with PVP (polyvinyl pyrrolidone) and the DNA extracted from the recovered cells. Generally only 30% of the cells in a soil are recovered in this way. Alternatively, DNA can be extracted directly from the soil by lysis of the cells *in situ* and purification again with PVP and other procedures such as hydroxy-apatite chromatography or caesium chloride density gradient centrifugation. Hybridization of the purified DNA to rRNA probes will reveal the presence of various classes of micro-organisms in the original sample. More specific probes will detect bacteria with certain traits such as the detection of the herbicide-degrading bacterium *Pseudomonas cepacia* in soils. The procedure can be used with all types of probes.

A neat variation of hybridization to DNA extracted from the environment is to do the experiment in reverse, that is to label the extracted DNA and to hybridize this to reference (probe) DNA on the filter. If the latter is chromosomal DNA, hybridization will provide identification of bacteria in the original sample. This 'reverse sample genome probing' has been used to identify 20 different sulphate-reducing bacteria in samples from Canadian oilfields (Voordouw *et al.*, 1991).

The PCR reaction (see Chapter 5) can be used to enhance the sensitivity of probing nucleic acids from the environment (Atlas, 1991). DNA extracted from soils can be amplified using primers specific for certain bacteria or their genes and the amplified DNA used in the hybridization. The increase in sensitivity is about 1000-fold. In one application, 100 ml of water is filtered, the cells lysed on the filter and *E. coli* sequences amplified with primers directed to the β-galactosidase and β-glucuronidase genes after processing of the filter to avoid excessive primer-binding. The same filter is then used for the hybridization. Single cells can be detected in the 100 ml samples.

7.8.3   Non-radioactive probes

Perhaps the biggest impetus to nucleic acid probe technology has come from the development of non-radioactive methods for the detection of the hybrid. At first these alternatives were clumsy and lacked the sensitivity of ^{32}phosphate, but many are now simple to use and have equivalent detection limits to radionuclides. Their big advantages are extended shelf-lives and safety compared with ^{32}P. The principal procedures are listed in Table 7.4. Direct methods incorporate a label into the DNA either by incorporating the modified nucleotide during a polymerase reaction or by attaching it to the end in the case of oligonucleotide probes. These enzyme-linked modified bases are then detected in the hybrid by a colorimetric reaction. Indirect labelling

Table 7.4   Some non-radioactive labelling and detection systems for nucleic acid hybridizations

Reporter group	Label	Detection
(a) Direct labels for hybridization probes		
	Alkaline phosphatase	Enzyme activity using X-phosphate
	Horseradish peroxidase	Chemiluminescence and 'autoradiography'
	Fluorescein tetramethylrhodamine	Fluorescent microscopy of whole cells
(b) Indirect detection of labelled probes		
Biotin	Streptavidin alkaline phosphatase	Enzyme activity using X-phosphate
Digoxigenin	Anti-dig antibody with alkaline phosphatase	Enzyme activity using X-phosphate

systems involve the incorporation of a modified nucleotide into the probe (reporter group). This nucleotide is then recognized by an intermediate such as an antibody to which the label is attached. The label is then detected as above. Chemiluminescent reactions are becoming popular because the hybridized filters can be exposed to film rather like an autoradiograph to provide a permanent record of the experiment.

The use of fluorescent probes for determinative microscopy is an interesting recent innovation. Following fixation most micro-organisms are permeable to short oligonucleotides. rRNA probes labelled with fluorescent dyes are used to hybridize with the rRNA target *in situ* and the cells incorporate sufficient dye to be visualized microscopically. In this way individual cells can be detected in tissues, body fluids, water or soil samples (Amann *et al.*, 1992).

Probe technology in clinical and environmental microbiology is currently expanding at an impressive rate. Improvements in detection limits promised by the PCR reaction and automation and quantification of the hybridized product will certainly extend the usefulness of the procedure. Indeed in years to come, we might simply reach for our probe to effect an identification rather than inoculating a range of media. However as a word of caution, a positive reaction only infers the presence of the organism. Panic may result if, for example, inactive vaccine cells were confused with the cells of a virulent pathogen.

REFERENCES

Amann, R., Ludwig, W. and Schleifer, K.-H. (1992). Identification and *in situ* detection of individual bacterial cells, *FEMS Microbiology Letters*, **100**, 45–50.

Atlas, R.M. (1991). Environmental applications of the polymerase chain reaction. *American Society for Microbiology News*, **57**, 630–632

Barrow, I. and Feltham, K.A. (Eds.). (1993). *Cowan and Steel's Manual for the Identification of Medical Bacteria*, 3rd edition. Cambridge University Press, Cambridge.

Bascomb, S., Lapage, S.P., Curtis, M.A. and Willcox, W.R. (1973). Identification of bacteria by computer: identification of reference strains. *Journal of General Microbiology*, **77**, 291–315.

Dybowski, W. and Franklin, D.A. (1968). Conditional probability and the identification of bacteria: a pilot study, *Journal of General Microbiology*, **54**, 215–229.

Gutteridge, C.S. (1987). Characterization of micro-organisms by pyrolysis mass-spectrometry, *Methods in Microbiology*, **19**, 227–272.

Krieg, N.R. (Ed.). (1984). *Bergey's Manual of Systematic Bacteriology, Vol. 1*, Williams and Wilkins, Baltimore.

Lapage, S.P., Bascomb, S., Willcox, W.R. and Curtis, M.A. (1973). Identification of bacteria by computer: general aspects and perspectives, *Journal of General Microbiology*, **77**, 273–290.

Macfie, H.J.H., Gutteridge, C.S. and Norris, J.R. (1978). Use of canonical variates analysis in differentiation of bacteria by pyrolysis gas-liquid chromatography, *Journal of General Microbiology*, **104**, 67–74.

Pankhurst, R.J. (1978). *Biological Identification; the Principles and Practice of Identification Methods in Biology*, Edward Arnold, London.

Payne, L.C. (1963). Towards medical automation, *World Medical Electronics*, **2**, 6–11.

Priest, F.G. and Williams, S.T. (1993). Computer-assisted identification, in *Handbook of New Bacterial Systematics* (M. Goodfellow and A.G. O'Donnell, Eds.), pp. 361–381, Academic Press, London.

Sayler, G.S. and Layton, A.C. (1990). Environmental applications of gene probes, *Annual Review of Microbiology*, **44**, 625–648.

Shute, L.A., Berkeley, R.C.W., Norris, J.R. and Gutteridge, C.S. (1985). Pyrolysis mass spectrometry in bacterial systematics, in *Chemical Methods in Bacterial Systematics* (M. Goodfellow and D. Minnikin, Eds.), pp. 95–114. Academic Press, London.

Smibert, R.M. and Krieg, N.R. (1981). General characterization, in *Manual of Methods for General Bacteriology* (P. Gerhardt, Ed.), pp. 409–443, American Society for Microbiology, Washington, DC.

Sneath, P.H.A. (1978). Identification of microorganisms, in *Essays in Microbiology* (J.R. Norris and M.H. Richmond, Eds.), pp. 10/1–10/32. Wiley, Chichester.

Stahl, D.A. and Amann, R. (1991). Development and application of nucleic acid probes, in *Nucleic Acid Techniques in Bacterial Systematics* (M. Goodfellow and E. Stackebrandt, Eds.), pp. 205–248, Wiley, Chichester.

Voordouw, G., Voordouw, J.K. Karkhoff-Schweizer, R.R. *et al.* (1991). Reverse sample genome probing, a new technique for identification of bacteria in environmental samples by DNA hybridization and its application to identification of sulfate-reducing bacteria in oilfields, *Applied and Environmental Microbiology*, **57**, 3070–3078.

Ward, D.M., Bateson, M.M., Weller, R. and Ruff-Roberts, A.L. (1992). Ribosomal RNA analysis of micro-organisms as they occur in nature, *Advances in Microbial Ecology*, **12**, 219–286.

Willcox, W.R., Lapage, S.P. and Holmes, B. (1980). A review of numerical methods in bacterial identification, *Antonie van Leeuwenhoek*, **46**, 233–299.

Williams, S.T., Goodfellow, M. Wellington, E.H.M. *et al.* (1983). A probability matrix for identification of some streptomycetes, *Journal of General Microbiology*, **129**, 1815–1830.

CHAPTER 8

Interactions between taxonomy and allied disciplines

It is now relevant to consider some of the wider reasons for taxonomic studies. Of course, it is possible to argue the need for taxonomy from the philosophical standpoint, i.e. the acquisition of knowledge to satisfy the curiosity of taxonomists. This would have been an acceptable premise for the Victorian era when science was largely the hobby of the rich, but, alas, such freedom of action has long since waned. The current attitude is one of constant justification for resources. Therefore, science needs to have appeal in order to wrest monies away from the ever-decreasing coffers. Fortunately, there are many areas of science in which a knowledge of bacterial taxonomy is essential or extremely useful. These include ecology, pathology, genetics and biotechnology, and culture collections.

8.1 ECOLOGY

The role of micro-organisms in the natural environment has gained importance, particularly in relation to the survival of pathogens, nutrient cycling and pollution. There is a need to recognize certain components of a microflora, e.g. coliforms, that are indicators of insanitary conditions. With the current popularity of biotechnology, there is interest in locating groups of organisms with special functions, such as antibiotic production. This necessitates detailed knowledge of the biology of the organisms. Unfortunately the classification of bacteria from the natural environment has been neglected (Table 8.1). Usually, ecologists have not been interested in taxonomy, and *vice versa*. Therefore, it is pertinent to determine the criteria for choosing meaningful representative strains from mixed microbial populations, such as occur in nature. Emphasis has been placed on general methods designed for medical bacteria rather than developing specialist systems suitable for ecology. Ecologists

Table 8.1 Some poorly defined or heterogeneous taxa associated with the natural environment

Bacterial taxa	Leaf surfaces	Rhizosphere soil	Freshwater	Estuarine environment	Marine environment
Achromobacter spp.	+	+	+		+
Acinetobacter spp.	+	+	+	+	+
Aerobacter spp.	+		+		
Arthrobacter spp.		+			
Brevibacterium spp.	+	+			
Chromobacterium spp.	+	+	+	+	+
Corynebacterium spp.	+	+			
Flavobacterium spp.	+	+	+	+	+
Flexibacter spp.	+	+	+	+	+
Lactobacillus spp.	+	+			
Micrococcus spp.	+	+	+	+	+
Myxococcus spp.	+				
Paracolobactrum aerogenoides	+				
Pseudomonas carnea	+				
P. epiphytica	+				
P. ichthyodermis					
P. incognita	+				
P. schuylkilleonsis	+				
Sarcina spp.	+				
Staphylococcus spp.	+	+	+	+	+

have identified isolates by means of conventional (medical) diagnostic schemes, although it is evident that the natural environment, including soil, leaf surfaces and water, is populated by bacteria that are likely to be different from those encountered in medical laboratories. Conversely, taxonomists, who expressed interest in the species composition of natural microbial populations, have tended to examine isolates of dubious ecological significance. The end result is often chaotic, with many environmental isolates identified as poorly defined or heterogeneous taxa. Even with the use of more modern procedures, such as numerical taxonomy, progress has been hampered by the extreme difficulty of identifying phena. Again, scientists have resorted to use of conventional schemes such as Cowan's (1974) identification tables or the older editions of *Bergey's Manual of Determinative Bacteriology*. With these, several distinct phena may be identified to the same species on the basis of a few 'key' features. Thus, these methods have been largely unsuccessful in identifying the wide variety of bacteria occurring in natural habitats. It is apparent that '... the identification of an unknown organism presupposes that others like it have been previously examined, compared with more organisms and named' (Gray, 1969).

If taxonomy is inadequate, it is very difficult to make meaningful judgements about the patterns in microfloras. Ecologists have tended to emphasize a few subjectively chosen 'key' morphological characteristics of microfloras. Colony pigmentation is a readily observable trait; and this has been repeatedly highlighted. Thus, populations of leaf surface (phylloplane) and marine bacteria contain large proportions of yellow-pigmented types. In contrast, the soil contains a microflora that produces mostly cream or white colonies. It is tempting to infer that the phylloplane and marine microfloras are more closely related to each other than to the soil. Yet, many of the yellow-pigmented bacteria have been traditionally identified as *Flavobacterium*, which became a heterogeneous taxon based largely upon pigmentation. Consequently, with reports that the phylloplane and the sea contain large populations of '*Flavobacterium*', it is easy to realize why scientists may link the two vastly different habitats. With further study, it has been determined that marine micro-organisms comprise a diverse range of unique taxa (Austin, 1988). However in the coastal environment, a high proportion of the organisms may be derived from the land, constituting a 'wash-in' component. Indeed, the term 'resident' has been coined to distinguish the indigenous component of the microflora, capable of multiplying in the natural environment, from organisms that only have a transient phase. However, from the techniques used to study microbial populations in the natural environment, it is difficult to distinguish between the resident and transient components.

Similarly, the soil contains some yellow flavobacteria, chromogenically related to a particular component of the phylloplane microflora. It is speculative whether or not the leaf surface bacteria originated in the soil, or *vice versa*. However, it is now realized that seed-borne bacteria may colonize both the phylloplane and rhizoplane (root surface); the latter being in intimate association with the soil.

Sweeping ecological generalizations have followed the acquisition of minimal data. For example, it was considered that many of the bacteria isolated from the phylloplane occurred on a wide range of plant species (Last and Deighton, 1965). Yet, it was unwise to put too much weight on distribution patterns because of the unsatisfactory state of classification of phylloplane bacteria. Nevertheless detailed investigation showed that some bacteria, notably *Beijerinckia and Spirillum* spp., were isolated only from tropical plants and appear to have restricted host ranges. The success of this study by Ruinen (1961) stemmed from the detailed examination of a few taxa, noted for their nitrogen metabolism, rather than a broad spectrum approach, in which sweeping generalizations are made about all aspects of a natural microbial population. The latter may be unconvincing, even with more modern approaches.

8.1.1 The application of numerical taxonomy to ecological studies

It may be envisaged that the introduction of numerical taxonomy led to welcome advances in ecology. Certainly, this technique has been used with bacterial populations derived from freshwater, estuarine and marine samples, soil, root surfaces, leaves and leaf litter (reviewed by Goodfellow and Dickinson, 1985). The first application was published by Brisbane and Rovira (1961), who studied 43 'rhizosphere' (the zone around roots) isolates and 21 reference cultures for 20 micromorphological, biochemical, degradative, antibiotic sensitivity and ecological tests. The study can hardly be considered as exhaustive, and the outcome of the analysis was not particularly helpful. Thus, 31 of the environmental isolates were recovered in one cluster, defined at the 85% similarity level, whereas the reference cultures clustered separately. This problem has re-occurred in many subsequent numerical taxonomy studies, involving environmental isolates, with reference cultures contributing little to the identification of phena. Part of the problem may reflect the difficulties of choosing meaningful reference strains, but there is the additional possibility that organisms, maintained in artificial laboratory conditions, lose or modify some of the characteristics associated with freshly isolated bacteria. Nevertheless, after this auspicious start by Brisbane and Rovira, the numerical studies continued, using more isolates and larger batteries of tests. Unfortunately, many of these studies have not capitalized on the outcome of the analyses to aid the understanding of ecology. It would appear that in many cases large collections of strains have been characterized solely for the purpose of carrying out numerical taxonomy studies.

When taxonomy and ecology have interacted the outcome has often been helpful. For example, the ecological study of soil bacteria by Lowe and Gray (1973a, b) was only possible because of the initial taxonomy (Lowe and Gray, 1972). Lowe and Gray examined 209 isolates and 21 reference cultures for 179 unit characters. Following analyses by the simple matching and Jaccard coefficients with clustering by single linkage, 180 of the isolates together with eight marker strains were recovered in seven clusters on the basis of overall similarity. Representative cultures were subsequently used in growth and competitive interaction experiments. In a larger study, Hissett and Gray (1974) examined 400 cultures, isolated from woodland litter and soil horizons, using numerical analyses. Although it is difficult from their publication to determine the precise techniques used, the end result showed that of 19 clusters, seven phena contained soil isolates, seven comprised litter isolates, and five phena contained both litter and soil organisms. Unfortunately the phena were not identified, but were somehow allocated to 'approximate taxonomic positions'. The study enabled soil isolates to be generally distinguished from those recovered from litter, and underlined differences between populations in these contrasting habitats. One of us (B.A.) used numerical taxonomy methods to study 621 isolates recovered from the phylloplane of perennial ryegrass. Again, difficulty ensued with attempts to equate phena with recognized taxa. However, representative strains were carefully chosen for ecological work, aimed at studying antagonistic interactions between phylloplane bacteria and a plant pathogenic fungus. The most convincing antagonist was *Pseudomonas fluorescens*, a taxon that appeared on the leaves only during the warmer summer months when the fungal pathogen, *Drechslera dictyoides*, was most troublesome.

It may be concluded that numerical taxonomy has successfully enabled some bacteria from heterogeneous populations to be grouped together in homogeneous clusters on the basis of shared characters. Once these clusters have been defined then *a posteriori* weighting of the group characters for use in diagnostic schemes becomes a valid proposition. However, only a few of the studies, involving isolates of ecological importance, have attempted to devise diagnostic aids suitable for identifying fresh isolates. This is a serious shortcoming of the other investigations insofar as comparisons between studies are therefore more difficult to achieve. Other shortcomings, in common with many of the numerical taxonomy studies, include problems with the choice of stable reference cultures, the selection of 'good' characters, and the difficulties with negative correlations.

The inclusion of suitable reference cultures seems, at times, an impossible task, and in the studies outlined above few were recovered in the clusters. For example, Lowe and Gray (1972) reported that only 38% of the named strains clustered with the environmental isolates. Even this low proportion is good compared with many other studies. Apart from the choice of suitable reference cultures, the selection of adequate characters has been a serious weakness of numerical taxonomy studies. The nature and number of the tests is subject to the discretion of each worker, yet these tests constitute a vital aspect of any investigation. Ideally a wide range of characters should be used from which an understanding of the overall properties of the organism may be achieved. It could be argued that ecologically based tests would be advantageous when studying organisms recovered from the natural environment. However, a danger would then exist that the work could not be compared with studies of other habitats. Consequently, it would be impossible to deduce whether or not the microflora of habitat A was distinct from habitat B. It is better, therefore, to use tests that can be directly applied or modified to suit the study of organisms from a wide range of habitats. The problems appertaining to test error and reproducibility have been discussed (see Chapter 2), yet, in many cases, numerical taxonomy studies of environmental isolates were based on small numbers of characters (e.g. Brisbane and Rovira, 1961), and errors could have caused severe distortions in the overall clustering of the isolates. Moreover, negative correlation (see Chapter 2) is particularly troublesome, insofar as many environmental isolates, e.g. *Flexibacter* spp., are quite unreactive in conventional testing regimes, and consequently could be clustered together on the basis of characters they did not possess. Thus, a false impression of homogeneity between dissimilar organisms would occur.

8.1.2 Ribosomal RNA analyses in ecological studies

It will be clear from the above account of the traditional approaches to taxonomy as applied to ecological studies that there were major difficulties with the identification and delineation of natural bacterial populations. Moreover, a point of immense importance is that numerical identification and other classical approaches require that the organisms must first be sampled from the environment, cultivated and subsequently identified. Since less than 20% of bacteria are known, and the vast majority cannot be cultured, we are examining at best the ecology of a minority of micro-organisms. In 1955, van Niel stated 'There are still many microbes whose natural functions are as yet unknown since they have not been encountered in elective culture' and almost 30 years later, Atlas emphasized the same problem: '...bacteriologists who rely on cultural methods to identify species, face the problem of selectivity and thus the inevitable underestimation of community diversity'. A major technological leap forward was needed to raise ecology from the limitations of cultural methods to the precise description of micro-organisms in the environment. Microbial ecologists were quick to realize that rRNA analyses could help them achieve this objective (Ward *et al.*, 1992).

Small and large subunit rRNAs can be used in two ways in the study of microbial ecology. Nucleic acids can be extracted from the environment and characterized. When the sequences are analysed in the context of the established database, this gives an indication of the range of micro-organisms in that particular ecological niche. It includes both dead and living micro-organisms but, more importantly, it also includes both culturable and non-culturable micro-organisms. These data provide descriptions of microbial community composition. Alternatively, probes can be directed to specific sequences (signatures) in the rRNA molecules in a sample from the environment. Bacteria can be identified from the sample, and in this way the autecology of micro-organisms can be studied irrespective of whether they can be cultured or not.

Isolation of the nucleic acid from the marine or terrestrial environment is generally achieved by physical disruption of the bacteria or enzymic

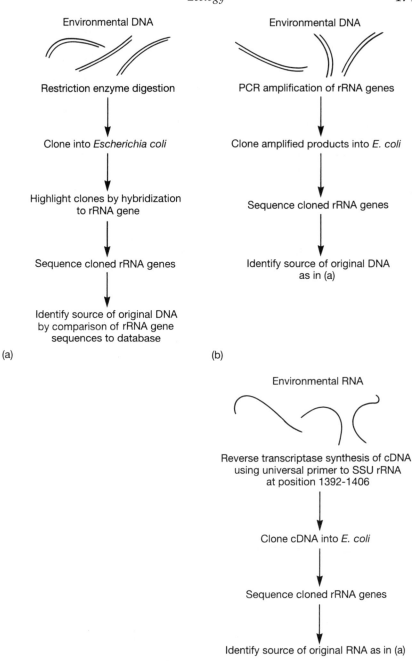

Fig. 8.1 Three approaches to the sequencing of rRNA genes from environmental samples for studies of population composition. (a) Shotgun cloning; (b) polymerase chain reaction; (c) cDNA synthesis.

lysis. The nucleic acids can then be further purified if necessary before analysis (see Chapter 7).

For studies of population composition, several approaches to the cloning and sequencing of rRNA from the environmental sample have been designed (Fig. 8.1). An original method was to shotgun clone the purified DNA into an appropriate vector and transform it into *E. coli*. Clones containing rRNA genes were identified by hybridization to an *E. coli* rRNA gene (about 0.2% of clones). These were then sequenced and the sequences compared with database sequences for identification of the cloned rRNA (Fig 8.1(a)). More recently, the polymerase chain reaction (PCR) has been used in this context. Primers to the ends (either 'universal' or designed for a particular group of organisms) are used to amplify rRNA sequences from the environmental sample. These fragments are then sequenced as described in Chapter 3 either directly or after cloning into *E. coli* (Fig. 8.1(b)). As before, the organisms are identified by comparison of the sequences to the rRNA database. Complications can arise in this method from imperfect primer annealing with recovery of sequences from non-target groups due to mismatching. Thirdly, rRNA sequences can be retrieved from environmental samples using reverse transcriptase (Fig 8.1(c)). A selective primer to the 3'-end of the 16S rRNA gene permits selective synthesis of a cDNA direct from the rRNA in the sample. This cDNA can then be cloned and sequenced. The advantage of this method is that it allows quantitative analysis of the rRNAs since it exploits the natural levels of rRNA in the sample. Finally, direct sequencing of rRNA in the sample can be performed, although this is restricted to the 5S molecule because of difficulties involved in processing the larger molecules.

The first three methods mentioned above result in large numbers of clones of rRNA sequences, many of which may be the same. They can be compared by sequence analysis of variable domains only (requiring only one sequencing reaction per clone), RFLP analysis using 4 bp cutting enzymes to provide numerous small fragments for comparison or on the basis of a single dideoxy reaction (T-lane analysis) in which only the distribution of one base in the sequence is compared. Subsequently, differing clones are analysed in more detail for classification and identification of the parental organisms.

Probe-based analyses use oligonucleotides directed to portions of the rRNA molecules depending on the specificity required. Highly variable regions are targeted for species-specific probes, less variable regions for genus-specific probes and even more highly conserved regions for kingdom-specific probes. For example in *Streptomyces*, a genus-specific probe was prepared centred on nucleotide 929 and the flanking region of the 16S rRNA gene. This probe hybridized with 77 streptomycetes and various environmental isolates but no related, non-streptomycete bacteria. Moreover, regions 158–203 of the 16S rRNA sequence and 1518–1645 of the 23S rRNA gene were shown to have a high potential to define species, although comparison of sequences with a database rather than probes was necessary to identify strains (Stackebrandt *et al.*, 1991).

Ecological applications of rRNA-based analyses have been numerous and are reviewed in detail by Ward *et al.* (1992). Here we will give a few examples to illustrate the power of the approach. Bacterioplankton in the oceans contribute significantly to biomass and biogeochemical activity, but the characterization of these communities was severely hampered because the majority of the organisms, as detected by direct counts, could not be cultured. Britschgi and Giovannoni (1991) approached the identification of bacterioplankton by using PCR (Fig. 8.1(b)) to construct and analyse a library of 16S rDNA genes isolated from Sargasso Sea bacterioplankton genomic DNA. Primers complementary to bacterial rRNA genes were used to accommodate a wide range of organisms. Of the 51 clones examined, RFLP analysis showed at least five classes of genes; SAR83 (nine clones), SAR92 (three clones), SAR175 (three clones), SAR132 (one clone) and the remaining 36 clones with no sites for the restriction enzymes used. Some of these 36 clones were allocated to two groups by specific hybridization with oligonucleotide probes; SAR11

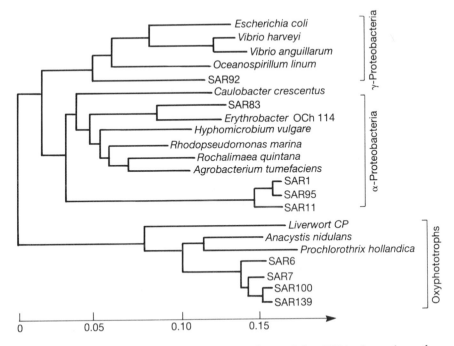

Fig. 8.2 Phylogenetic tree showing relationships of the rDNA clones from the Sargasso Sea to representative cultivated species (from Britschgi and Giovannoni, 1991, with permission).

(nine clones) and the *Synechococcus* group (SAR100, SAR152 and SAR139). The remaining 24 clones could not be easily identified (47%). The sequences of several unique genes were determined fully and a tree prepared to show the classification of some of these bacterioplankton organisms (Fig. 8.2). Some were unknown members of the proteobacteria and others were a new lineage of cyanobacteria. This initial study supports the view that the majority of bacterioplankton has not been cultured and the large number of unidentified sequences indicates that there is considerable undiscovered diversity in the oceans. However, these early studies cannot yet be definitive since the sequences recovered might represent taxa in culture but not yet sequenced.

Various symbioses between bacteria and animals are being explored using phylogenetic approaches. The teredinid molluscs (shipworms) are widely distributed throughout the oceans and are responsible for the destruction of wooden structures and the recycling of organic carbon from cellulose in the marine environment. The molluscs depend on a bacterial symbiosis to enable them to survive on a diet of wood. Large numbers of Gram-negative bacteria localized within a specialized tissue in the gill of the mollusc actively degrade the wood cellulose and also fix nitrogen, thus supplying their host with dietary carbon and nitrogen. Distel *et al.* (1991) have sequenced the 16S rRNA genes from these symbionts of *Lyrodus pedicellatus* and placed the bacterium in the

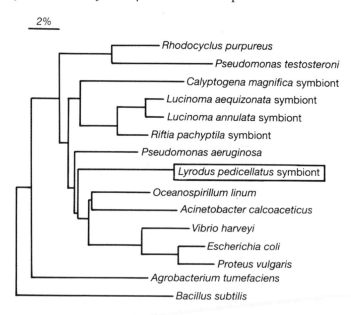

Fig. 8.3 Phylogenetic tree showing the relationship of the nitrogen-fixing, cellulolytic bacterium of the shipworm *(Lyrodus pedicellatus)* with selected reference species from the Proteobacteria (from Distel *et al.,* 1991, with permission).

gamma-3 subdivision of the proteobacteria (Fig. 8.3). The bacterium is distinct from all other known bacterial genera. In order to certify that this bacterium was indeed the major symbiont of *L. pedicellatus*, a specific oligonucleotide probe was prepared, labelled with a fluorescent dye and hybridized with the endosymbionts *in situ*. High hybridization to endosymbionts in the symbiont-containing vesicles of the gill showed that the initial teredinid isolates were representative of the gill populations.

These two examples illustrate the power of phylogenetic-based analyses in microbial ecology and evolution. No doubt, as the techniques become more widespread far-reaching insights into all aspects of microbial ecology will be achieved.

8.2 PATHOLOGY

The role of micro-organisms in disease processes was one of the principal founding interests in microbiology. It is apt to recall Koch's postulates, which essentially considered the need to isolate a pathogen in pure culture, to confirm its pathogenicity, and then to re-isolate the same organism from the infection. The first and last stages demand the need to recognize an organism. Yet 100 years after Koch, the ability of diagnosticians to identify organisms remains questionable. Perhaps, the problem concerns the large volume of work handled by most diagnostic laboratories, and, therefore, the need to spend only the minimal amount of time and resources on each specimen. This situation is complicated by the overwhelming need to kill the pathogen in the infected host, rather than to carry out extensive characterizations. Consequently, much attention is focused on antibiotic sensitivity testing. The outcome is that new or interesting pathogens may be missed. Moreover, epidemiology studies may be hampered. The taxonomy of pathogens receives effort only when new patterns of disease become apparent. The case of legionnaire's disease became well documented in the popular press after an outbreak in the USA, which resulted in numerous deaths. The causal agent was eventually isolated, and described in a new genus, as *Legionella pneumophila*. Subsequent investigation has revealed that legionnaire's disease is not a new condition, but has been regarded in the past as 'atypical pneumonia'. Even after considerable publicity, legionnaire's disease was initially mis-diagnosed as influenza (type B) during a serious outbreak in Stafford, UK, which occurred in spring 1985. Similarly, *Erwinia herbicola* (now *Enterobacter agglomerans*) was announced as an emerging pathogen of human beings, yet a culture had been deposited in a culture collection 40 years previously, albeit under the name *Chromobacterium typhiflavum*. Therefore, it was necessary for good observation to pinpoint the presence of a hitherto undescribed human pathogen. Even *Aeromonas hydrophila*, an ubiquitous freshwater inhabitant, has been recognized to be a serious pathogen. Previously, isolates had been merely labelled as atypical coliforms. With the general awareness of the presence of these serious pathogens, meaningful epidemiological investigations have been possible.

The saga of unrecognized pathogens is not confined to human medicine. The taxonomy of plant pathogenic xanthomonads was complicated by the system of naming new species after the host plant, without detailed characterization to confirm the unique position of the so-called 'new' organisms. Fortunately, detailed taxonomic work has led to the recognition of synonyms, and thus to a marked reduction in the total number of *Xanthomonas* species. For example, one of us (B.A.) recognized the high level of relatedness between reference cultures of *X. begoniae*, *X. campestris* and *X. hedera*, which did not warrant the presence of three separate species names. Indeed, this group of xanthomonads was subsequently determined to comprise a dominant component of the phylloplane microflora of healthy plants. Therefore, this habitat would form the likely reservoir for infection.

Veterinary medicine has also benefited from the attention of bacterial taxonomists. The speciation of coryneforms and mycobacteria has almost become a separate specialist branch of science. *Corynebacterium pyogenes* has been reclassified into the genus *Actinomyces*, as *Actinomyces pyogenes* largely on the basis of chemotaxonomy. Mycobacteriosis may appear to be a syndrome caused by a multiplicity of species, yet many of these may be reduced to synonyms of a few well-defined taxa. Perhaps the confusion has been caused by an abundance of reports based largely on histological examination of infected material. In fact, some diagnosticians have not bothered to isolate the offending pathogens; instead, diagnoses have resulted solely from the presence of acid-fast organisms in tissue samples. It is relevant in these cases to know how mycobacteria may be differentiated from nocardia, which also cause serious infections in animals. Perhaps the diagnosticians are gifted with considerable taxonomic intuition! Yet, the all-important control regimes may differ between mycobacteria and nocardia, and mis-diagnosis may lead to inappropriate treatment.

The blinkered approach to diagnosis often causes interesting pathogens to be missed. For example, *Listonella* (=*Vibrio*) *anguillarum* has long been regarded as the causal agent of vibriosis in marine fish. The disease may be described as a haemorrhagic septicaemia. Consequently, many such cases of disease have been attributed to vibriosis, particularly if the pathogen displays certain key characteristics, namely growth on thiosulphate citrate bile salt sucrose agar, and the presence of motile Gram-negative fermentative rods which produce catalase and oxidase. Many organisms were identified as *L. anguillarum*, before heterogeneity among the isolates became obvious. Initially, the taxon was divided into two biotypes, of which the latter became reclassified as *V. ordalii*. The relevance of this observation to fish pathology is that *bona fide* strains of *L. anguillarum* are more serious pathogens than *V. ordalii*. Moreover, this information has to be considered in the development of vaccines, which may necessitate the need for multivalent products. Other vibrios are now recognized as causal agents of fish diseases, including *V. alginolyticus*, *V. carchariae*, *V. cholerae*, and *V. vulnificus*. Undoubtedly, as more attention is given to this area of pathology, other 'new' pathogens will be recognized.

Sometimes, impressions may be gained of serious epidemics affecting wide geographic areas. In one such case, in fish, streptococcicosis was identified in Japan, followed by outbreaks in South Africa and the USA. However, the causal agent was not fully characterized until recently when *Enterococcus seriolicida* was described. However, it is unclear whether all the outbreaks of streptococcicosis may be attributed to *E. seriolicida* or if other species are involved. This is of paramount importance from the epidemiological viewpoint.

The difficulties, discussed above, have resulted largely from incomplete taxonomic knowledge. At one extreme, there is haste to describe new taxa. Conversely, poor taxonomy may lead to the failure to recognize patterns in disease outbreaks. There is a tendency to follow certain lines of work, regardless of whether or not they are suited to the conditions. Rapid diagnosis of disease is essential for swift corrective action. Serology is particularly suited for such speed, yet the specificity of the reactions has rarely been considered. It is appropriate to consider the hundreds of *Salmonella* 'species' included in the earlier editions of *Bergey's Manual of Determinative Bacteriology*. These 'species' were created solely as a result of serology. Fortunately, these taxa have been reduced in status, to serological variants of a few well-defined species. Such finely divided groups are of value only for epidemiology, where it is important to trace the source and spread of infections, but have little relation to pathogenicity where the clone (in a population genetic sense) has more relevance.

Rapid diagnosis can also be achieved using DNA probe technology. Oligonucleotide probes can be used to detect and identify pathogens in pathological samples such as tissues, faecal or urine samples. Probes have even been used to detect uncultured pathogens in tissue samples, namely the aetiological agent of bacillary angiomatosis, a disease which is characterized by a proliferation of small blood vessels in the skin of immunocompromised hosts (Relman *et al.*, 1990). The possibilities for diagnostic DNA probes for other 'difficult to culture' pathogens, such as the slow-growing mycobacteria which cause leprosy and tuberculosis in man, are very exciting and represent a good example of taxonomy leading the developments in diagnostic clinical microbiology.

8.3   GENETICS AND MOLECULAR BIOLOGY

Until recently, the interaction between molecular biology and taxonomy was entirely unidirectional, involving the adoption of molecular and chemical methods by taxonomists to improve their classifications and provide superior identification systems. Taxonomy has been largely ignored by molecular geneticists. This could be interpreted as flexibility of the systematist, who is prepared to explore new techniques, and the relatively narrow outlook of the molecular geneticist, who generally aims to analyse one particular strain in detail. This was understandable during the development of bacterial genetics, since exploitable systems of gene exchange were only available for a few organisms. However, the situation is now changing, and the molecular geneticist might benefit greatly from reading some of the new systematics. One reason for this is the development of gene cloning techniques. It is now possible, theoretically at least, to introduce genes from a Gram-negative bacterium into any other Gram-negative bacterium and maintain them stably. This is possible through the use of the broad-host range plasmids of incompatibility group P as vectors for cloned genes (Heinemann, 1991). For Gram-positive bacteria, such vectors were not available until recently and plasmids with more restricted host ranges were used. In the case of *Bacillus subtilis*, a genetically well-characterized organism, indigenous plasmids with selectable markers, such as antibiotic resistance, initially proved difficult to detect, and it was not until 1977 that plasmids from *Staphylococcus aureus* were found to be suitable for cloning in *Bacillus* owing to their stable replication and maintenance in this host. Had the close molecular relationship between these Gram-positive cocci and bacilli, which has now been demonstrated by rRNA cataloguing and sequencing, been acknowledged earlier, it is likely that the *S. aureus* plasmids would have been exploited in *B. subtilis* genetics at an earlier stage.

Moreover, the close phylogeny of the bacilli and lactobacilli makes it more understandable that the new cloning vectors based on pAMβ 1 and relatives, which were originally isolated in lactic acid bacteria, transfer to and replicate in bacilli. These have a broader host range and are more stable than the *S. aureus* plasmids. Similarly the transposon, Tn917, a transposable DNA element providing resistance to erythromycin and used for transposon mutagenesis and the construction of gene fusions, was originally isolated from *Enterococcus* but has been used in other members of the low G+C group of Gram-positive bacteria including several bacilli and *Listeria*. Thus, an understanding of the classification of bacteria, particularly the cladistic relationships, can indicate to the geneticist groups of closely related bacteria among which plasmids, phage and transposons might be exchanged and stably maintained.

In a similar vein, there are barriers to heterologous gene expression. Genes from Gram-positive bacteria are usually transcribed and translated in *E. coli*, but Gram-negative genes are not normally expressed in *B. subtilis* and other Gram-positive eubacteria. Streptomycetes seem to express only actinomycete genes. Again, a sound knowledge of the molecular relationships between bacteria should indicate hosts for gene cloning that readily express heterologous genes and can be used in particular situations, such as lactobacilli in silage production or pseudomonads for hydrocarbon degradation. It seems likely, therefore, that as molecular biology is applied to more and varied bacteria, a comprehensive, phylogenetic classification of the bacteria will be an invaluable aid to the molecular biologists developing new hosts for genetic analyses.

8.4   BIOTECHNOLOGY

Biotechnology, the commercial exploitation of biology, relies heavily on microbiology and the microbial systematist is finding his skills are increasingly sought. The demand for novel pharmaceutical agents, be they antibiotics, antitumour or antiviral agents, enzymes or enzyme inhibitors is great. The list of valuable primary and secondary metabolites is now very long indeed and continues to grow as novel bacteria are isolated and cultured. Appreciation by the biotechnology companies of the hidden biodiversity in the microbial world and its offer of new and valuable metabolites has resulted in the systematist being in demand.

8.4.1   Isolation of novel bacteria

A major limitation to the development of screening programmes for new metabolites with pharmacological properties is the isolation of novel bacteria from the environment. It is argued that new microorganisms will provide new metabolites, but how does one go about cultivating new strains and how does one recognize novelty? Initially, microbiologists sampled exotic habitats ranging from insect intestines to the cold seas of Antarctica in the search for unusual bacteria, and this was reasonably effective. However, if the normal media and incubation conditions are used, common organisms, such as endospore forming rods, enteric bacteria and pseudomonads flourish during the isolation procedures and overgrow the rarer organisms. Even if the source material was derived from some exotic habitat, it is likely that common organisms will be isolated. In effect, typical isolation media and conditions give rise to typical bacteria, whatever the source.

One successful approach to the isolation of unusual micro-organisms exploited data from numerical taxonomy studies in a very effective way (Williams *et al.*, 1984). Initially, an exhaustive numerical taxonomic analysis of *Streptomyces* and related organisms provided a comprehensive data matrix comprising many poorly represented and uncommon taxa as well as the more common ones, together with their characteristics. The data were in the form of a frequency matrix providing characteristics of taxa (see Table 2.8). This showed the carbon source utilization patterns of taxa, their resistance profiles to antibiotics and other inhibitors of growth, temperature and pH ranges for growth and numerous other features. In addition it provided information on antibiotic production.

From this information, unusual taxa could be chosen which produced unusual antibiotics. Selective media were then formulated for these taxa, often containing three or more antibiotics against which strains of the selected taxon were known to be resistant, or comprising an unusual carbon source that could be used only by the selected organisms. Media were thus designed that would repress the more common organisms but permit the growth of selected unusual taxa. In this way, Williams and his colleagues were able to sample common habitats and isolate rare and unusual streptomycetes that were subsequently screened for the synthesis of novel antibiotics.

A second example relates to the isolation and identification of insect pathogenic bacteria with potential as biological control agents. As mentioned in Chapter 3, some *Bacillus sphaericus* strains are toxic to mosquito larvae, but the distribution of toxicity among strains of this rather diffuse species was until recently unclear. DNA reassociation studies and complementary numerical taxonomy analysis revealed several homology groups, with all the mosquito pathogens assigned to DNA homology group IIA. Scrutiny of the results of the numerical taxonomy indicated that members of this taxon were resistant to streptomycin and could use adenine as a source of carbon. A selective medium based on this knowledge has been useful for the isolation of insect pathogenic *B. sphaericus* strains from the environment. Thus, many more strains with increased pathogenicity or different host ranges could be readily isolated. Of course, this approach, whether used for streptomycetes or bacilli, only helps to isolate more examples of species for which we already have strains. It does not permit the isolation of unknown micro-organisms; a much more difficult task.

Screening programmes also need to guard against the continuous re-isolation of strains of existing taxa. When screening for new antibiotics how does one exclude the thousands of strains already screened and kept or discarded? One way is to use novel and targeted isolation conditions, as mentioned above. Alternatively some rapid method of identification can be useful to identify colonies quickly and efficiently. Thus, colonies can be identified and discarded or shown not to match to an existing database and retained. Two methods of rapid identification are used in screening programmes for this purpose. Py-MS has been popular because of the speed with which identifications can be achieved (see Chapter 4). Alternatively, DNA probes are used; these have the advantage that non-culturable organisms can be detected and identified in samples (Bull *et al.*, 1992).

8.4.2 Traditional technologies

Microbial taxonomists also contribute to the efficient operation of traditional biotechnological industries, such as the food and dairy industries, brewing and other industrial fermentations for enzymes or antibiotics. Any large-scale fermentation or manufacture of a food is susceptible to contamination during production, or spoilage of a packaged product. Such contamination may be detrimental to the process and reduce yields, may affect the flavour, efficiency or appearance of the product or may be deleterious to health, if consumed. Rapid detection and identification of spoilage organisms is, therefore, vitally important to the efficient manufacture of these products.

Computerized identification procedures (as outlined in Chapter 7) are proving to be useful in this context. A typical problem could involve the microbial spoilage of beer. A numerical taxonomy study of Gram-positive cocci in beer and breweries revealed *Pediococcus damnosus* and *P. pentosaceus* strains, various staphylococci and micrococci as the dominating flora (Lawrence and Priest, 1981). However, of these bacteria only *P. damnosus* strains are able to grow in and spoil beer; the low pH, and presence of hop iso-acids inhibit the other bacteria, which presumably gain access to the beer from the air, raw materials and human contact. Isolation of staphylococci or micrococci from beer during normal quality control procedures could lead to panic decisions to pasteurize and recycle the beer or even destroy it, if the bacteria were thought to be the spoilage organism *P. damnosus*. However, by operating a rapid identification scheme in the laboratory on a microcomputer, the quality control scientist can quickly and accurately identify the offending bacteria, and, if they are harmless micrococci which are unable to grow in and spoil the product, the beer can be processed as normal. Obviously the source of the bacteria would have to be located and eliminated, but there would be no immediate threat to the product. This is just one of many situations in which a rapid, reliable and predictive identification service can save substantial sums of money and lead to more efficient and productive manufacturing processes.

Predictably, there have been numerous publications describing DNA probes for the detection and identification of pathogens in foods. For example numerous probes for *Listeria monocytogenes* (the causative agent of listeriosis) have been developed, partly because traditional cultural methods for its isolation and identification take from a few days to several weeks. In one study, Golsteyn Thomas *et al.* (1991) cultured food samples in an enrichment broth and used primers complementary to the listeriolysin O gene for PCR amplification. Since only clinical strains of *L. monocytogenes* contain this gene, it is a useful marker for specific identification. The products are analysed by agarose gel electrophoresis and are specific for *L. monocytogenes*. These PCR assays could detect as little as 1 cell per 100 ml of milk or 1 cell per gram of minced meat in less than 3 days.

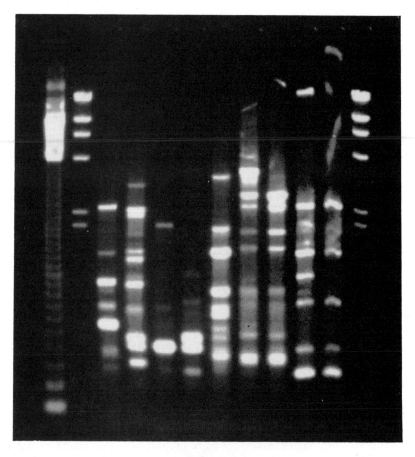

Fig. 8.4   RAPDs typing of some strains of *Neisseria*. This RAPD reaction was carried out with a single primer and 45 PCR cycles. The lanes are, from left to right; 1, 123 bp ladder size marker; 2, Lambda *Hind*III digest size marker; 3, *N. gonorrhoeae*; 4, *N. flavescens*; 5, *N. lactamica*; 6, *N. sicca*; 7, *N. polysaccharae*; 8, *N. meningitidis* serogroup B; 9, *N. meningitidis* serogroup B; 10, *N. meningitidis* serogroup A; 11, *N. meningitidis* serogroup B; 12, Lambda *Hind*III size markers (photograph courtesy of M. Maiden).

8.4.3 Typing of strains

The biotechnology industry often needs to identify one particular strain and discriminate it from all others. Production strains are usually covered by patents, but how does one identify one's own particular patented strain? This is a 'typing' problem, rather similar to paternity cases in humans, and PCR technology has again come into its own. In the randomly amplified polymorphic DNAs (RAPDs) approach, several (about six) short (about 10 base) random primers are used for PCR amplification of chromosomal DNA. The products are separated on an agarose gel by electrophoresis and provide a 'fingerprint' of the strain represented by numerous bands in the gel. It is not yet clear how effective this approach will be for typing bacteria at the individual strain level because the population genetics of micro-organisms is so poorly understood and sequence variation between species is largely unknown. A typical profile is shown in Fig. 8.4 for the same *Neisseria* strains as those examined by RFLP analysis in Fig. 3.9.

8.5 CULTURE COLLECTIONS

Culture collections are the microbiologist's botanical or zoological garden, in which representative micro-organisms are preserved and maintained. They provide an indispensable service to microbiologists through the provision of well-characterized, authenticated strains of viruses, bacteria, fungi, protozoa and, more recently animal and cell lines. Many countries have, or are introducing, these 'service' collections, which supply micro-organisms to universities, schools, hospitals, research centres and industry. Some countries, such as the USA and Germany, have single centralized culture collections (American Type Culture Collection and Deutsche Sammlung von Microorganismen und Cell, respectively), but in the UK several specialized national culture collections hold more than 30 000 strains, ranging from viruses to animal cell lines. These collections are represented nationally by the UK Federation for Culture Collections and internationally by the European Culture Collection Organization and the World Federation of Culture Collections, which provides overall coordination.

In addition to the distribution of well-characterized, authenticated strains, culture collections are also centres of expertise in the fields of preservation, classification and identification of bacteria. Preservation is particularly important with the current interest in biodiversity. Many culture collections offer comprehensive catalogues with detailed histories and applications of the strains they hold, and identification services for unknown organisms. Frequently, the catalogues will be computerized, so that screens for strains with specific activities can be readily achieved. Expansion of computerized services is being achieved through the MIRCEN (Microbiological Resource Centres), MSDN (Microbial Strain Data Network) and MINE (Microbial Information Network for Europe) programs which offer, in different ways, access to computerized databases of micro-organisms and their properties.

The recent and expanding interest in biotechnology is heavily dependent on culture collections, firstly as a source of useful organisms and also for the deposition of cultures. If a process involving a micro-organism is patented, a culture of that strain must be deposited in a culture collection that is an accepted International Depository Authority. Moreover, culture collections offer other services to industry, including the preservation of production strains, strain selection programmes and, as mentioned previously, identification of unknown organisms. Culture collections are indispensable to microbiology, and it is encouraging that international cooperation is leading to the maintenance of a wider range of micro-organisms and cell lines in collections. Moreover, computerized data handling and on-line access to the mass of information housed in these collections will make them increasingly useful (see Kirsop, 1983; Malik and Claus, 1987).

REFERENCES

Austin, B. (1988). *Marine Microbiology*, Cambridge University Press, Cambridge.

Brisbane, P.G. and Rovira, A.D. (1961). A comparison of methods for classifying rhizosphere bacteria, *Journal of General Microbiology*, **26**, 379–392.

Britschgi, T.B. and Giovannoni, S.J. (1991). Phylogenetic analysis of a natural marine bacterioplankton population by rRNA gene cloning and sequencing. *Applied and Environmental Microbiology*, **57**, 1707–1713.

Bull, A.T., Goodfellow, M. and Slater, H. (1992). Biodiversity as a source of innovation in biotechnology, *Annual Review of Microbiology*, **46**, 219–252.

Cowan, S.T. (1974). *Cowan and Steel's Manual for the Identification of Medical Bacteria*, 2nd edition. Cambridge University Press, Cambridge.

Distel, D.L., De Long, E.F. and Waterbury, J.B. (1991). Phylogenetic characterization and in situ localization of the bacterial symbiont of shipworms (Teredinidae: Bivalvia) by using 16S rRNA sequence analysis and oligodeoxynucleotide probe hybridization. *Applied and Environmental Microbiology*, **57**, 2376–2382.

Golsteyn Thomas, E., King, R.K., Burchak, J. and Gannon, V.P. (1991). Sensitive and specific detection of *Listeria monocytogenes* in milk and ground beef with the polymerase chain reaction. *Applied and Environmental Microbiology*, **57**, 2576–2580.

Goodfellow, M. and Dickinson, C.H. (1985). Delineation and description of microbial populations using numerical methods, in *Computer-Assisted Bacterial Systematics* (M. Goodfellow, D. Jones and F.G. Priest, Eds.), pp. 165–225, Academic Press, London.

Gray, T.R.G. (1969). The identification of soil bacteria, in *The Soil Ecosystem* (J.G. Sheals, Ed.), pp. 73–82. The Systematics Association Publication No. 8.

Heinemann, J.A. (1991). Genetics of gene transfer between species, *Trends in Genetics*, **7**, 181–185.

Hissett, R. and Gray, T.R.G. (1974). Bacterial populations of litter and soil in a deciduous woodland. I. Qualitative studies, *Revue d'ecologie et de Biologie du Sol*, **10**, 495–508.

Kirsop, B. (1983). Culture collections–their services to biotechnology, *Trends in Biotechnology*, **1**, 4–8.

Last, F.T. and Deighton, F.C. (1965). The non-parasitic microflora on the surface of living leaves, *Transactions of the British Mycological Society*, **48**, 83–99.

Lawrence, D.R. and Priest, F.G. (1981). Identification of brewery cocci, in *Proceedings of the European Brewery Convention, Copenhagen*, pp. 217–227, Elsevier, Amsterdam.

Lowe, W.E. and Gray, T.R.G. (1972). Ecological studies on coccoid bacteria in a pine forest soil. I. Classification, *Soil Biology and Biochemistry*, **4**, 459–467 .

Lowe, W.E. and Gray, T.R.G. (1973a). Ecological studies on coccoid bacteria in a pine forest soil. II. Growth of bacteria inoculated into soil, *Soil Biology and Biochemistry*, **5**, 449–452.

Lowe, W.E. and Gray, T.R.G. (1973b). Ecological studies on coccoid bacteria in a pine forest soil. III. Competition interactions between bacterial strains in soil, *Soil Biology and Biochemistry*, **5**, 463–472.

Malik, K.A. and Claus, D. (1987). Bacterial culture collections: their importance to biotechnology and microbiology, *Biotechnology and Genetic Engineering Reviews*, **5**, 137–197.

Relman, D.A. Loutit, J.S., Schmidt, T.M. *et al.* (1990). The agent of bacillary angiomatosis: an approach to the identification of uncultured pathogens, *New England Journal of Medicine*, **323**, 1573–1580.

Ruinen, J. (1961). The phyllosphere. I. An ecologically neglected milieu, *Plant and Soil*, **15**, 81–109.

Stackebrandt, E., Witt, D., Kemmerling, C. *et al.* (1991). Designation of streptomycete 16S and 23S rRNA-based target regions for oligonucleotide probes, *Applied and Environmental Microbiology*, **57**, 1468–1477.

Ward, D.M., Bateson, M.M., Weller, R. and Ruff-Roberts, A.L. (1992). Ribosomal RNA analysis of microorganisms as they occur in nature, *Advances in Microbial Ecology*, **12**, 219–286.

Williams, S.T., Goodfellow, M. and Vickers, J.C. (1984). New microbes from old habitats?, *Symposium of the Society for General Microbiology*, **36(II)**, 219–256.

CHAPTER 9

Conclusions and outlook

Since its origins as a science in the nineteenth century, bacterial taxonomy has enjoyed a phase of rising interest with a concomitant increase in knowledge, particularly in the years since 1945. New techniques, such as chemotaxonomy and numerical taxonomy, have been developed. The value of these approaches is attested by their widespread use throughout the scientific community. Thus, within only a few decades, bacterial taxonomy had progressed from a comparatively mystical craft to a respectable exacting science. What developments seem likely to follow in the future? Predicting the future is a notoriously unreliable business, but within the realms of bacterial taxonomy there are some developments which should reach fruition within the next few years. For example, increasingly sophisticated techniques are being devised for the analysis of nucleic acids, many of which have taxonomic value. Clearly, the message is that advances in taxonomy depend upon developments in allied disciplines. It is certain that the current interest in chemotaxonomy and molecular phylogenies will continue. However, care needs to be given to the degree of emphasis placed on such information in the formulation of classifications. At present, many Gram-positive taxa are defined largely on the basis of chemotaxonomic characters, and it remains to be seen whether or not this is sound policy. It would appear to be more beneficial for many such characters to be incorporated in numerical taxonomy studies; otherwise, undue weight will be given to what is, after all, often limited information on the biology of the organisms under study.

Classifications based largely on DNA:DNA hybridizations escape from this criticism because the complete genetic information of the bacteria under study is being used. Moreover, as stressed in Chapter 3, the composition of nucleic acids is totally unaffected by the environment, and any organism may be compared with any other under standard conditions. Thus nucleic acid analyses are the only source of standardized data for the delineation of taxa, although arguments concerning the level of genetic heterogeneity allowable within a species still flourish. On similar grounds, rRNA analyses seem to be the most appropriate criteria for the delineation of higher ranked taxa, such as genera and families. It is important that classifications are consistent with those based on the phenotype and it is heartening that, in general, this is the case. Where there are persistent differences between the phenetic and phylogenetic classifications, it is not clear which discipline will 'win', although the requirement for rapid, simple identification will probably emphasize the phenetic classification – at least in the near future.

There is no doubt that molecular phylogeny is the single most stimulating and exciting development in microbial systematics. Among its many successes are: an insight into early evolution on this planet; an indication of the origins of the organelles of eukaryotic cells; provision of a route to the classification and identification of 'non-culturable' micro-organisms; and the development of identification probes for a wide range of applications. The enthusiasm with which molecular systematics has been adopted by microbiologists has left the scientific community with severe data handling problems and the urgent requirements are for consensus on treeing algorithms and extensive deposition of unpublished sequences in databases to avoid duplication of effort. To this end, the international RNA database project should be helpful.

With the publication of the Approved Lists of Bacterial Names in 1980 and its more recent updates, bacterial nomenclature has become a reasonably orderly process and, at last, identification is being treated with the importance it deserves. The conventional approaches of morphological and biochemical tests and the paraphernalia of dichotomous keys and/or diagnostic tables are gradually being replaced by commercial kits and computerized databases for clinically important bacteria; but for most non-medical organisms only the traditional schemes are available. Fortunately, the availability of computer hardware and software is enabling those scientists interested in disciplines other than medical microbiology to generate probabilistic identification matrices for various bacterial groups. Such matrices will be included in future editions of *Bergey's Manual of Systematic Bacteriology* for general usage. Although we foresee routine identification of bacteria largely relying on phenotypic testing in the near future, it has been suggested that in the next century we will simply reach for a bottle of probe for identification in much the same way as we might mix together two chemicals to identify them.

Serology has been revolutionized by the use of highly specific monoclonal antibodies and their incorporation into sensitive serological techniques, such as the enzyme linked immunosorbant assay (ELISA) and immunohistochemistry techniques. Reliable rapid diagnoses are possible with these systems, and there is not a need for expensive equipment. Indeed, they can often be used in the field to examine diseased animals or plants for the presence of specific organisms. Techniques such as ELISA are likely to be the subject of great commercial interest in the quest for diagnostic kits for the mass market. Not only will this interest improve the diagnosis of disease but the techniques should also aid industry in recognizing particular strains. For example, these could have industrial significance in the production of valuable compounds, such as antibiotics and enzymes. In short, the message is that bacterial taxonomy is not a subject doomed to museums. Instead, taxonomy is an exciting branch of science with considerable promise for the future. We await, with considerable interest, the next important landmark in the development of bacterial taxonomy. Such activities of tomorrow will result from the students of today. Therefore, the kindling of interest in microbial systematics among our readers may generate the ideas to develop and even replace the currently popular molecular systematics.

Glossary

Additive coding A system of coding quantitative features into binary code, which retains the scale of the character.

Algorithm A set of instructions for a series of calculations.

Alignment The pairing of two homologous sequences for the purpose of identifying insertions and deletions.

Allele Alternative form of a gene at a locus.

Allozyme An allelic form of an enzyme.

Analogy Similarity by convergent evolution, but not by common evolutionary ancestry.

A posteriori **(L.)** 'From what comes after', i.e. inductively.

Approved Lists of Bacterial Names A list of validly published bacterial names that appeared in the January 1980 issue of the *International Journal of Systematic Bacteriology*. Any names not included in these or subsequent updates are not nomenclaturally valid.

A priori **(L.)** 'From what is before', i.e. presumptively.

Back mutation A mutation which reverses the effect of a previous mutation and restores the original nucleotide sequence.

Binary character A character existing in two states, as positive (+) or negative (−).

Binomial name Owing its origin to the Swedish naturalist Linnaeus, the name of a species comprises two words, the first being the genus name, and the second the species name, i.e. specific epithet.

Canonical variates analysis See Multiple discrimination analysis.

cDNA (complementary DNA) DNA that has been prepared from RNA using reverse transcriptase.

Centroid A point representing the centre of a cluster.

Centrotype The nearest OTU to the centroid.

Character From the taxonomic standpoint, the term refers to any measurable property.

Chemotaxonomy Taxonomy based upon chemical analyses of whole cells or subcellular components.

Chemotype A representative strain for a set of chemical characteristics.

Clade (a) A taxon comprising a single species and all its descendants representing a monophyletic branch in an evolutionary tree.
(b) A subgroup of organisms from among a larger group sharing a common ancestor not shared by the other organisms in the group.

Cladistic Refers to the branching pattern that describes the pathway of ancestry of a group of organisms.

Cladogram A tree-like diagram that represents the evolutionary pathway of a group of organisms.

Classification The process of ordering organisms into groups.

Clone A population of identical cells.

Cluster analysis The mathematical determination of groups (clusters) from a distance or similarity matrix.

Coefficient   A measure used to calculate the similarity or distance between two OTUs.

Combinatio nova   (comb. nov.) New combination, applies to a name which may result from, for example, transfer of a species from one genus to another resulting from improvements in its taxonomy.

Congruence   The level of agreement between two taxonomic methods.

Convergence   The independent evolution of similar genetic or phenotypic traits.

Cophenetic correlation   Measure of the accuracy with which a dendrogram represents a similarity matrix.

Dendrogram   A tree-like diagram resulting from cluster analyses, which expresses relationships between OTUs.

Diagnosis   A brief description of a taxon, permitting its differentiation from all others. The word is often used in the context of disease.

Dichotomous (diagnostic) keys   Identification keys in which the sequential answering of questions ultimately provides an identification.

Discriminant analysis   *See* Multiple discrimination analysis.

Dissimilarity   The exact opposite of similarity.

Distance   (a) A measure of dissimilarity between OTUs.
(b) Genetic: A measure of the number of nucleotide substitutions per nucleotide site between two homologous DNA sequences.

Divergence index   An indication of sequence divergence established by dividing DNA reassociation measured at a stringent temperature by that at the optimal temperature.

Domain   A taxon, ranked above Kingdom, which accommodates the major phyletic lines; Archaea, Bacteria and Eucaryota.

Equal weight   Appertains to the notion that in numerical classification, all characters (tests) are of the same importance.

Euclidean distance   The measurement of distance between OTUs in terms of their coordinates on right-angled (Cartesian) axes.

Genus novum   (gen. nov.) New genus.

Hierarchy   An organization with ranks, graded one above the other.

Holotype   A type strain reported by the original author.

Hypothetical median organism   These are so-called 'average' organisms and are representative of clusters of OTUs. The term was coined by Liston and coworkers in 1963.

Hypothetical taxonomic units   OTUs with theoretical characteristics used in the construction of phylogenetic trees.

Identification   The process by which organisms are assigned to taxonomic groups, which have been previously defined as a result of classification.

Illegitimate name   A name not published in accordance with the *International Code of Nomenclature of Bacteria*.

Incongruence   The opposite of congruence, i.e. the level of disagreement between two taxonomies.

International Code of Nomenclature of Bacteria   A formal code, originally mooted at the Second International Congress of Microbiology in 1936, which seeks to ensure the presence of stable, meaningful bacterial names.

Isochronic   In the present context, the term refers to individuals within a group evolving at a similar rate.

Judicial Commission   Approved by the Second International Congress for Microbiology in 1936 to issue formal nomenclatural 'Opinions', on request.

Lectotype   If a subsequent author designates one of the original author's strains as a type culture, it is referred to as the lectotype.

Legitimate name   A name published in accordance with the *International Code of Nomenclature of Bacteria*.

Matrix (similarity or dissimilarity) In numerical taxonomy, this is a triangular array of similarity or distance estimates, which permits the comparison of an OTU with any other listed in the data set.

Metric An analogous term for dissimilarity measurements.

Minimal evolution The concept that evolution proceeds along the shortest possible pathway with the fewest number of steps (see Parsimony).

Molecular clock The rate at which mutations accumulate in a nucleic acid.

Monophyletic group A phylogenetic group which shares a common ancestor.

Monothetic A group, membership of which is dependent on the presence or absence of a few invariant characters.

Monotype Where an original author described only one strain, which was not specifically regarded as the holotype, it is regarded as the monotype.

Multiple discrimination analysis Similar to Principal component analysis, but the axes represent the greatest variability of the means of the different taxa. Used for identification.

Multistate characters Characters/tests which occur in more than one state, e.g. tests which may be scored as negative, weakly, moderately, or strongly positive and could be recorded numerically as 0, 1, 2 or 3, respectively.

Natural (phenetic) classification Natural, in this context, refers to classifications based upon as many aspects of the biology of the organisms as possible.

Neotype A culture of the original author's batch designated as a type culture by a subsequent author.

Nomenclatural type *See* Type strain.

Nomen conservandum Conserved name – as voiced through an Opinion of the Judicial Commission. Conserved names must be used instead of synonyms.

Nomen rejiciendum Rejected name – as voiced through an Opinion of the Judicial Commission.

Nomenclature The process of allocating names to groups of organisms.

Numerical taxonomy The arrangement of organisms into groups on the basis of their overall characteristics, as a result of numerical methods.

Operational taxonomic unit (OTU) Within the realms of bacteriology, refers to the bacterial isolate/strain which is the subject of study.

'Opinion' A formal view on a matter of nomenclature expressed by the Judicial Commission.

Ordination methods Mathematical methods which reduce the number of dimensions in a taxonomic space to two or three, such that distance between OTUs can be represented.

Outgroup A species that is the least related to the others in a group of species. Used for comparative purposes.

Paralogy Sequence similarity between the descendants of a duplicated ancestral gene.

Parsimony The theory that evolution proceeds via the fewest number of steps.

Pattern difference As discussed by Professor P.H.A. Sneath, the term applies to one of the two components of the total difference between OTUs. Pattern difference (D_p) may be expressed as: $D_p = [2\sqrt{(bc)}]/(a+b+c+d)$ (see Chapter 2 for interpretation of symbols).

Phenetic classification A classification based on the overall properties of the organisms, as they are perceived at present.

Phenon (pl. phena) A group defined on the basis of high overall similarity of all of its members.

Phenotype Appertains to observable properties, without regard to ancestry.

Phylogeny The study of the evolution/ancestry of organisms.

Polymerase chain reaction (PCR) A technique which enables multiple copies of a DNA fragment to be generated by enzymic amplification of a target DNA sequence.

Polymorphism   A variable DNA sequence, one that can exist in several different but related forms.

Polyphyletic   Descended from different ancestors.

Polythetic   Refers to a group characterized by a high number of common features (characters). Membership of the group does not require the presence or absence of any particular attribute. In this context, phena are polythetic.

Principal components analysis   An ordination method in which the first principal component is derived as a dimension representing the greatest variability among the data and plotted against a second dimension which accounts for the next greatest variability.

Principal coordinate analysis   An ordination method similar to Principal components analysis but applicable to taxonomic data not necessarily based on Euclidean distance. If the distances are Euclidean, it becomes the same as Principal components analysis.

Priority   Within the terms of the *Bacteriological Code of Nomenclature*, priority is given to the first valid publication of a name.

Pseudogene   A non-functional nucleotide sequence which has similarity to a functional gene but within which the biological information has become scrambled.

Randomly amplified polymorphic DNAs (RAPDs)   Cell 'fingerprints' obtained by electrophoresis of DNA fragments generated by PCR amplification of chromosomal DNA using random primers.

Restriction endonuclease   An enzyme that cuts DNA molecules at specific nucleotide sequences.

Restriction fragment length polymorphism (RFLP)   Differences in sizes of restriction fragments resulting from a DNA molecule digested by a restriction enzyme.

Scatter diagram   A representation of the position of OTUs in a two-dimensional taxonomic space.

Second site reversion   A second mutation that reverses the effect of a previous mutation but without restoring the original nucleotide sequence.

Similarity   In the context of numerical taxonomy, the term refers to the level of resemblance between OTUs.

Small subunit RNA (SSU RNA)   The 16S and 18S ribosomal RNA molecules of the prokaryotic and eukaryotic cells, respectively.

Speciation (cladogenesis)   The process by which a new species arises.

Species   (a) The basic taxonomic unit, which should consist of highly related isolates.

(b) A group of interbreeding individuals.

Species nova   (sp. nov.) New species.

Specific epithet   The second word of a binomial name, indicating the identity of a species.

Stable RNA   RNA molecules, such as ribosomal and transfer RNA but not messenger RNA which are not subject to rapid turnover in the cell.

Systematics   The study of the diversity and relationship among organisms.

Taxometrics   *See* Numerical taxonomy.

Taxon (pl. taxa)   A group of individuals of any rank.

Taxonomic map   The two- or three-dimensional arrays showing the ordering of OTUs as a result of ordination methods.

Taxonomy   The theory of classification, nomenclature and identification.

Taxospecies   A group of isolates with a mutually high phenetic similarity.

Test error/reproducibility   A term first coined by Professor P.H.A. Sneath for the mistakes which may creep into the recording of characters. Unfortunately, the definition of positive and negative responses of some tests is vague.

Thermal binding index   *See* Divergence index.

Transformed cladistics   Classifications based on maximum parsimony but with no recourse to evolutionary theories.

Transition   A point mutation that results in a purine being replaced by another purine or pyrimidine.

Transversion   A point mutation in which a pyrimidine is replaced by a purine or *vice versa.*

Type (reference) strain   An organism culture which has been designated as a permanent representative of a taxon. Type strains will be deposited in culture collections.

Ultrametric property   The fact that a cladogram will be the same as a phenogram/dendrogram if evolutionary rates are constant and convergence minimal.

Unit character   A taxonomic character (test) of two or more states, which cannot be subdivided logically except for changes in the method of coding.

Vigour difference   As discussed by Professor P.H.A. Sneath, this term applies to the differences in growth rates between OTUs.
Mathematically, the vigour difference (D_v) may be expressed as: $D_v = (c-b)/(a + b + c + d)$ (see Chapter 2 for the meaning of the symbols).

Appendix A: Classification of the bacteria – the conventional approach

This scheme has been based on data from *Bergey's Manual of Systematic Bacteriology*, Vol. 1 to 4, and the *Approved Lists of Bacterial Names* (1980) and their supplements. Some anomalies are apparent. For example, the genus *Moraxella* has been included in two separate families, i.e. Branhamaceae and Moraxellaceae.

* Refers to the type genus (see Chapter 6). The type species for each genus is given in parentheses.

It should be emphasized that, unlike botanical and zoological classifications, bacteria do not all fit into a convenient hierarchical arrangement, culminating in a single hyper-group at the tip of a pyramid. This point will become abundantly clear. Nevertheless for convenience, the bacteria are considered divided as follows:

Kingdom:	Procaryotae
Division:	Gracilicutes (Gram-negative bacteria)
Class:	Scotobacteria
Class:	Anoxyphotobacteria
Class:	Oxyphotobacteria
Class:	Proteobacteria
Division:	Firmicutes (Gram-positive bacteria)
Class:	Firmibacteria
Class:	Thallobacteria
Division:	Tenericutes (Bacteria which lack rigid cell walls)
Class:	Mollicutes
Kingdom:	Archaebacteria
Division:	Mendosicutes (Bacteria with unusual cell walls)
Class:	Archaeobacteria

Within this framework all bacteria should be placed. However to simplify matters, the bacteria have been grouped according to certain traits:

Phototrophic bacteria

Order: Prochlorales

Family: Prochloraceae

Genus: *Prochloron (P. didemni)*

Family: Prochlorotrichaceae

Genus: *Prochlorothrix (P. hollandica)*

Order: Rhodospirillales

Family: Chromatiaceae

Genus: *Chromatium (C. okenii)*
Lamprobacter (L. modestohalophilus)
Thiocystis (T. violacea)
Thiopedia (T. rosea)

Amoebobacter (A. roseus)
Lamprocystis (L. roseopersicina)
Thiodictyon (T. elegans)
Thiospirillum (T. jenense)

Family: Chlorobiaceae

Genus: *Chlorobium (C. limicola)*
Pelodictyon (P. clathratiforme)

Chloroherpeton (C. thalassium)
Prosthecochloris (P. aestuarii)

Related genus: Heliobacterium (H. chlorum)

Family: Ectothiorhodospiraceae

Genus: *Ectothiorhodospira (E. mobilis)*

Family: Rhodospirillaceae

Genus: *Rhodospirillum (R. rubrum)*
Rhodocyclus (R. purpureus)
Rhodomicrobium (R. vannielii)
Rhodopseudomonas (R. palustris)

Rhodobacter (R. capsulatus)
Rhodoferax (R. fermentans)
Rhodopila (R. globiformis)

Related genus: Erythrobacter (E. longus)
Roseobacter (R. litoralis)

Lamprobacter (L. modestohalophilus)
Rubrivivax (R. gelatinosus)

Genera of uncertain affiliation:

Chloroflexus (C. aurantiacus)

Heliothrix (H. oregonensis)

Gliding bacteria

Order: Myxobacterales
  Family: Archangiaceae
    Genus: *Archangium (A. gephyra)
  Family: Cystobacteraceae
    Genus: *Cystobacter (C. fuscus)
           Stigmatella (S. aurantiaca)
           Melittangium (M. boletus)
  Family: Polyangiaceae
    Genus: *Polyangium (P. vitellinum)
           Nannocystis (N. exedens)
           Chondromyces (C. crocatus)

Order: Cytophagales
  Family: Cytophagaceae
    Genus: *Cytophaga (C. hutchinsonii)
           Flexithrix (F. dorotheae)
           Saprospira (S. grandis)
           Flexibacter (F. flexilis)
           Herpetosiphon (H. aurantiacus)
           Sporocytophaga (S. myxococcoides)
           Chitinophaga (C. pinensis)
           Sphingobacterium (S. spiritivorum)
  Related genus: Capnocytophaga (C. ochracea)
                 Microscilla (M. marina)
                 Thermonema (T. lapsum)
  Family: Flavobacteriaceae
    Genus: *Flavobacterium (F. aquatile)
  Family: Leucotrichaceae
    Genus: *Leucothrix (L. mucor)
           Thiothrix (T. nivea)
  Family: Simonsiellaceae
    Genus: *Simonsiella (S. muelleri)
           Alysiella (A. filiformis)

Order: Lysobacterales
  Family: Lysobacteraceae
    Genus: *Lysobacter (L. enzymogenes)

Order: Myxococcales
    Family: Myxococcaceae
        Genus: *Myxococcus (M. fulvus)*
    Families of uncertain affiliation:
    Family: Achromatiaceae
        Genus: *Achromatium (A. oxaliferum)*
    Family: Vitreoscillaceae
        Genus: *Vitreoscilla (V. beggiatoides)*
    Genera of uncertain affiliation:
        Filibacter (F. limicolor)

Angiococcus (A. disciformis)

Sheathed bacteria
    Family: Crenothrichaceae
        Genus: *Crenotrix (C. polyspora)*
    Genera of uncertain affiliation:
        Blastococcus (B. aggregatus)
        Leptothrix (L. ochracea)

Toxothrix (T. trichogenes)

Haliscomenobacter (H. hydrossis)
Sphaerotilus (S. natans)

Budding and appendaged bacteria

Order: Caulobacterales
    Family: Caulobacteraceae
        Genus: *Caulobacter (C. vibrioides)*
    Related genus: *Prosthecobacter (P. fusiformis)*

Order: Hyphomicrobiales
    Family: Hyphomicrobiaceae
        Genus: *Hyphomicrobium (H. vulgare)*
    Related genus: Hirschia (H. polymorpha)

Hyphomonas (H. polymorpha)
Pedomicrobium (P. ferrugineum)

Order: Planctomycetales
    Family: Planctomycetaceae
        Genus: *Planctomyces (P. bekefii)*
    Families of uncertain affiliation:

Pirellula (P. staleyi)

Family: Gallionellaceae
Genus: *Gallionella (G. ferruginea)
Family: Nevskiaceae
Genus: *Nevskia (N. ramosa)
Family: Pasteuriaceae
Genus: *Pasteuria (P. ramosa)
Genera of uncertain affiliation:

Ancalomicrobium (A. adetum) Angulomicrobium (A. tetraedrale)
Asticcacaulis (A. excentricus) Blastobacter (B. henricii)
Dichotomicrobium (D. thermohalophilum) Ensifer (E. adhaerens)
Filomicrobium (F. fusiforme) Gemmata (G. obscuriglobus)
Gemmobacter (G. aquatilis) Labrys (L. monachus)
Prosthecomicrobium (P. pneumaticum) Seliberia (S. stellata)
Stella (S. humoa) Verrucomicrobium (V. spinosum)

Aerobic/micro-aerophilic non-motile/motile helical/vibrioid Gram-negative bacteria

Order: Spirillales
Family: Spirillaceae
Genus: *Spirillum (S. volutans) Aquaspirillum (A. serpens)
Azospirillum (A. lipoferum) Bdellovibrio (B. bacteriovorus)
Herbaspirillum (H. seropedicae) Magnetospirillum (M. ryphiswaldense)
Prolinoborus (P. fasciculus) Oceanospirillum (O. linum)
Vampirovibrio (V. chlorellavorus)
Family: Campylobacteraceae
Genus: *Campylobacter (C. fetus) Arcobacter (A. nitrofigilis)
Related genus: Helicobacter (H. pylori)
Family of uncertain affiliation:
Family: Spirosomaceae
Genus: *Spirosoma (S. linguale) Cyclobacterium (C. marinus)
Flectobacillus (F. major) Runella (R. slithyformis)
Related genus: Ancylobacter (A. aquaticus) Brachyarcus (B. thiophilus)
Meniscus (M. glaucopsis) Pelosigma (P. cohnii)

Genus of uncertain affiliation:
    Halovibrio (H. variabilis)

Gram-negative aerobic rods and cocci

Family:	Acetobacteriaceae	
Genus:	*Acetobacter (A. aceti)	
Related genus:	Acidomonas (A. methanolica)	Gluconobacter (G. oxydans)
Family:	Azotobacteriaceae	
Genus:	*Azotobacter (A. chroococcum)	Azomonas (A. agilis)
	Azomonotrichon (A. macrocytogenes)	Azorhizophilus (A. paspali)
	Beijerinckia (B. indica)	Derxia (D. gummosa)
Family:	Brucellaceae	
Genus:	*Brucella (B. melitensis)	
Family:	Halobacteriaceae	
Genus:	*Halobacterium (H. salinarium)	Haloarcula (H. hispanica)
	Halococcus (H. morrhuae)	Haloferax (H. gibbonsii)
Family:	Legionellaceae	
Genus:	*Legionella (L. pneumophila)	Fluoribacter (F. bozemanae)
	Tatlockia (T. micdadei)	
Family:	Neisseriaceae	
Genus:	*Neisseria (N. gonorrhoeae)	Kingella (K. kingae)
	Mesophilobacter (M. marinus)	Morococcus (M. cerebrosus)
Family:	Branhamaceae	
Genus:	*Branhamella (B. catarrhalis)	
Family:	Moraxellaceae	Moraxella (M. lacunata)
Genus:	*Moraxella (M. lacunata)	
	Psychobacter (P. immobilis)	Acinetobacter (A. calcoaceticus)
Related genus:	Volcaniella (V. eurihalina)	
Family:	Rhizobiaceae	
Genus:	*Rhizobium (R. leguminosarum)	Agrobacterium (A. tumefaciens)
	Bradyrhizobium (B. japonicum)	Phyllobacterium (P. myrsinacearum)
Related genus:	Azorhizobium (A. caulinodans)	Ochrobacterium (O. anthropi)
	Sinorhizobium (S. fredii)	

Family:	Comamonadaceae
Genus:	*Comamonas (C. terrigena)	Acidovorax (A. facilis)
	Hydrogenophaga (H. pseudoflava)	Variovorax (V. paradoxus)
	Xylophilus (X. ampelinus)

Family:	Methylococcaceae
Genus:	*Methylococcus (M. capsulatus)	Methylomonas (M. methanica)
Related genus:	Methylobacillus (M. glycogenes)	Methylobacterium (M. organophilum)
	Methylophaga (M. marina)	Methylophilus (M. methylotrophus)
	Methylovorus (M. glucosotrophus)	Protomonas (P. extorquens)

Genus of uncertain affiliation:
	Agitococcus (A. lubricus)

Family:	Alcaligenaceae
Genus:	*Alcaligenes (A. faecalis)	Bordetella (B. pertussis)

Order:	Pseudomonadales
Family:	Pseudomonadaceae
Genus:	*Pseudomonas (P. aeruginosa)	Frateuria (F. aurantia)
	Xanthomonas (X. campestris)	Zoogloea (Z. ramigera)
Related genus:	Aminobacter (A. aminovorans)	Chromohalobacter (C. marismortui)
	Chryseomonas (C. polytricha)	Flavimonas (F. oryzihabitans)
	Rhizobacter (R. daucus)	Rhizomonas (R. ruberifaciens)
	Rugamonas (R. rubra)	Xylella (X. fastidiosa)

Genera of uncertain affiliation:
	Achromobacter (A. xylosoxidans)	Acidophilium (A. cryptum)
	Acidothermus (A. cellulolyticus)	Agromonas (A. oligotrophica)
	Alteromonas (A.macleodii)	Carboxydothermus (C. hydrogenoformans)
	Cupriavidus (C. necator)	Francisella (F. tularensis)
	Janthinobacterium (J. lividum)	Lampropedia (L. hyalina)
	Marinomonas (M. communis)	Natronobacterium (N. gregoryi)
	Natronococcus (N. occultus)	Oligella (O. urethralis)
	Paracoccus (P. denitrificans)	Phenylobacterium (P. immobile)
	Serpens (S. flexibilis)	Thermoleophilum (T. album)
	Thermomicrobium (T. roseum)	Thermus (T. aquaticus)

Order:
Family: Halobacteriales
 Halomonadaceae
Genus: *Halomonas (H. elongata)*
Related genus: Marinobacter (M. hydrocarbonoclasticus)

Xanthobacter (X. autotrophicus)

Weeksella (W. virosa)

Deleya (D. aesta)

Facultatively anaerobic Gram-negative rods

Family: Aeromonadaceae
Genus: *Aeromonas (A. hydrophila)
Family: Enterobacteriaceae
Genus: *Escherichia (E. coli)
 Budvicia (B. aquatica)
 Buttiauxella (B. agrestis)
 Citrobacter (C. freundii)
 Enterobacter (E. cloacae)
 Ewingella (E. americana)
 Klebsiella (K. pneumoniae)
 Koserella (K. trabulsii)
 Leminorella (L. grimontii)
 Moellerella (M. wisconsensis)
 Obesumbacterium (O. proteus)
 Pectobacterium (P. carotovorum)
 Proteus (P. vulgaris)
 Rahnella (R. aquatilis)
 Serratia (S. marcescens)
 Tatumella (T. ptyseos)
 Xenorhabdus (X. nematophilus)
 Yokenella (Y. regensburgei)

 Arsenophonus (A. nasoniae)

 Cedecea (C. davisae)
 Edwardsiella (E. tarda)
 Erwinia (E. amylovora)
 Hafnia (H. alvei)
 Kluyvera (K. ascorbata)
 Leclercia (L. adecarboxylata)
 Levinea (L. amalonatica)
 Morganella (M. morganii)
 Pantoea (P. agglomerans)
 Pragia (P. fontium)
 Providencia (P. alcalifaciens)
 Salmonella (S. choleraesuis)
 Shigella (S. dysenteriae)
 Trabulsiella (T. guamensis)
 Yersinia (Y. pestis)

Related genus: Saccharobacter (S. fermentatus)
Family: Pasteurellaceae
Genus: *Pasteurella (P. multocida)
 Haemophilus (H. influenzae)

 Actinobacillus (A. lignieresii)

Related genus: Gardnerella (G. vaginalis)
Family: Vibrionaceae
Genus: *Vibrio (V. cholerae)
  Enhydrobacter (E. aerosaccus)
  Photobacterium (P. phosphoreum)
  Shewanella (S. putrefaciens)
Related genus: Colwellia (C. psychroerythrus)
Family: Cardiobacteriaceae
Genus: *Cardiobacterium (C. hominis)
  Suttonella (S. indologenes)
Genera of uncertain affiliation:
  Calymmatobacterium (C. granulomatis)
  Eikenella (E. corrodens)
  Streptobacillus (S. moniliformis)

Taylorella (T. equigenitalis)

Allomonas (A. enterica)
Listonella (L. anguillarum)
Plesiomonas (P. shigelloides)

Micavibrio (M. admirandus)

Dichelobacter (D. nodosus)

Chromobacterium (C. violaceum)
Iodobacter (I. fluviatile)
Zymomonas (Z. mobilis)

Gram-negative anaerobic rods and cocci

Family: Bacteroidaceae
Genus: *Bacteroides (B. fragilis)
  Acetoanaerobium (A. noterae)
  Acidaminobacter (A. hydrogenoformans)
  Anaerovibrio (A. lipolytica)
  Centipeda (C. periodontii)
  Lachnospira (L. multiparis)
  Pectinatus (P. cerevisiiphilus)
  Propionigenium (P. modestum)
  Rikenella (R. microfusus)
  Selenomonas (S. sputigena)
  Succinivibrio (S. dextrinosolvens)
  Syntrophus (S. buswellii)
Related genus: Acetofilamentum (A. rigidum)
  Acetothermus (A. paucivorans)
  Capsularis (C. zoogleiformans)
  Formivibrio (F. citricus)

Acetivibrio (A. cellulolyticus)
Acetomicrobium (A. flavidum)
Anaerobiospirillum (A. succiniproducens)
Butyrivibrio (B. fibrisolvens)
Fusobacterium (F. nucleatum)
Leptotrichia (L. buccalis)
Pelobacter (P. acidigallici)
Propionispira (P. arboris)
Roseburia (R. cecicola)
Succinomonas (S. amylolytica)
Syntrophomonas (S. wolfei)
Wolinella (W. succinogenes)
Acetohalobium (A. arabaticum)
Bilophila (B. wadsworthia)
Fibrobacter (F. succinogenes)
Megamonas (M. herpermegas)

Mitsuokella (M. multiacidus) Mobiluncus (M. curtisii)
Porphyromonas (P. asaccharolytica) Ruminobacter (R. amylophilus)
Prevotella (P. melaninogenica) Acidaminococcus (A. fermentans)

Family: Veillonellaceae
Genus: *Veillonella (V. parvula)
Megasphaera (M. elsdenii)
Related genus: Genminger (G. formicilis)

Family: Haloanaerobiaceae
Genus: *Haloanaerobium (H. praevalens) Halobacteroides (H. halobius)
Haloincola (H. saccharolytica)

Family: Desulfurococcaceae
Genus: *Desulfurococcus (D. mucosus)
Related genus: Desulfobacter (D. postgatei) Desulfobacterium (D. indolicum)
Desulfobulbus (D. propionicus) Desulfococcus (D. multivorans)
Desulfohalobium (D. retbaense) Desulfomonas (D. pigra)
Desulfomonile (D. tiedjei) Desulfonema (D. limicola)
Desulfosarcina (D. variabilis) Desulfovibrio (D. desulfuricans)
Desulfurolobus (D. ambivalens) Desulfuromonas (D. acetoxidans)

Genera of uncertain affiliation:
Hyperthermus (H. butylicus) Syntrophobacter (S. wolinii)
Thermobacteroides (T. acetoethylicus)

Miscellaneous anaerobes

Genus: Llyobacter (L. polytropus) Malonomonas (M. rubra)
Oxalobacter (O. formigenes) Zymophilus (Z. raffinosivorans)

Miscellaneous thermophilic Gram-negative bacteria

Order: Thermoproteales
Family: Thermoproteaceae
Genus: *Thermoproteus (T. tenax) Thermophilum (T. pendens)
Aquifex (A. pyrophilus) Fervidobacterium (F. nodusum)
Related genus: Pyrobaculum (P. islandicum) Thermosipho (T. africanus)

Order: Thermococcales
Family: Thermococcaceae
Genus: *Thermococcus (T. celer) Pyrococcus (P. furiosus)

Order: Thermotogales
Family: Thermotogaceae
Genus: *Thermotoga (T. maritima)

Gram-negative chemolithotrophs

Ammonia or nitrite oxidizing bacteria
Family: Nitrobacteraceae
Genus: *Nitrobacter (N. winogradskyi) Nitrococcus (N. mobilis)
Nitrosococcus (N. nitrosus) Nitrosolobus (N. multiformis)
Nitrosomonas (N. europaea) Nitrosospira (N. briensis)
Nitrospina (N. gracilis)
Nitrospira (N. marina)

Related genus

Sulphur bacteria
Order: Sulfolobales
Family: Sulfolobaceae
Genus: *Sulfolobus (S. acidocaldarius) Stygiolobus (S. azoricus)
Related genus: Acidianus (A. infernus) Archaeoglobus (A. fulgidus)
Metallosphaera (M. sedula)

Genera of uncertain affiliation:
Macromonas (M. mobilis) Pyrodictium (P. occultum)
Thermothrix (T. thiopara) Thiobacillus (T. thioparus)
Thiomicrospira (T. pelophila) Thiosphaera (T. pantotropha)
Thiospira (T. winogradskyi) Thiovulum (T. majus)

Family: Thiocapsaceae
Genus: *Thiocapsa (T. roseopersicina)

Hydrogen bacteria
Genus: Acetonema (A. longum)
Calderobacterium (C. hydrogenophilum) Hydrogenobacter (H. thermophilus)

Methane bacteria

Order: Methanobacteriales
  Family: Methanobacteriaceae
  Genus: *Methanobacterium (M. formicicum)
    Methanobrevibacter (M. ruminatium)
    Methanosphaera (M. stadtmaniae)
  Related genus: Syntrophococcus (S. sucromutans)
    Hydrogenovibrio (H. marinus)

Order: Methanococcales
  Family: Methanococcaceae
  Genus: *Methanococcus (M. mazei)

Order: Methanomicrobiales
  Family: Methanomicrobiaceae
  Genus: *Methanomicrobium (M. mobiles)
    Methanolacinia (M. paynteri)
    Methanogenium (M. cariaci)
    Methanoculleus (M. bourgense)
    Methanospirillum (M. hungatii)

Families of uncertain affiliation:
  Family: Methanosarcinaceae
  Genus: *Methanosarcina (M. methanica)
    Methanolobus (M. tindarius)
    Methanosaeta (M. concilii)
    Methanococcoides (M. methyluteus)
    Methanothrix (M. soehngenii)
  Related genus: Methanohalophilus (M. mahii)
  Family: Methanoplanaceae
  Genus: *Methanoplanus (M. limicola)
  Family: Methanothermaceae
  Genus: *Methanothermus (M. fervidus)
  Family: Methanocorpusculaceae
  Genus: *Methanocorpusculum (M. parvum)
Genera of uncertain affiliation:
  Halomethanococcus (H. doii)
  Methanopyrus (M. kandleri)
  Methanohalobium (M. evestigatus)

Gram-positive cocci

Aerobic and / or facultatively anaerobic cocci
Order: Micrococcales
Family: Micrococcaceae
Genus: *Micrococcus (M. luteus)
Staphylococcus (S. aureus)
Related genus: Paracoccus (P. denitrificans)
Family: Planococcaceae
Genus: *Planococcus (P. citreus)
Family: Deinococcaceae
Genus: *Deinococcus (D. radiodurans)
Related genus: Deinobacter (D. grandis)
Genus of uncertain affiliation:
Salinicoccus (S. roseus)

Stomatococcus (S. mucilaginosus)

Marinococcus (M. albus)

Family of uncertain affiliation:
Family: Streptococcaceae
Genus: *Streptococcus (S. pyogenes)
Enterococcus (E. faecalis)
Leuconostoc (L. mesenteroides)
Trichococcus (T. flocculiformis)
Related genus: Lactococcus (L. lactis)
Vagococcus (V. fluvialis)
Genus of uncertain affiliation:
Saccharococcus (S. thermophilus)

Aerococcus (A. viridans)
Gemella (G. haemolysans)
Pediococcus (P. damnosus)

Melisococcus (M. pluton)

Anaerobic cocci
Family: Peptococcaceae
Genus: *Peptococcus (P. niger)
Peptostreptococcus (P. anaerobius)
Sarcina (S. ventriculi)

Coprococcus (C. eutactus)
Ruminococcus (R. flavefaciens)

Endospore-forming rods and cocci

Order: Bacillales
    Family: Bacillaceae
        Genus: *Bacillus (B. subtilis)*
                Sporolactobacillus (S. inulinus)
        Related genus: Amphibacillus (A. xylanus)
                            Desulfotomaculum (D. nigrificans)
                            Sporosarcina (S. ureae)

Order: Clostridiales
    Family: Clostridiaceae
        Genus: *Clostridium (C. butyricum)*
        Related genus: Sporohalobacter (S. lortetii)
    Family of uncertain affiliation:
                            Syntrophospora (S. bryantii)
    Family: Oscillospiraceae
        Genus: *Oscillospira (O. guilliermondii)*
    Genus of uncertain affiliation:
        Sulfobacillus (S. thermosulfidooxidans)

Gram-positive non-sporing rods

Family: Lactobacillaceae
    Genus: *Lactobacillus (L. delbreuckii)*
    Related genus: Alloiococus (A. otitis)
                Erysipelothrix (E. rhusiopathiae)
                            Brochothrix (B. thermosphacta)
                            Carnobacterium (C. divergens)
                            Listeria (L. monocytogenes)

Genera of uncertain affiliation:
        Falcivibrio (F. grandis)
        Thermoanaerobium (T. brockii)
                              Thermoanaerobacter (T. ethanolicus)

Actinomycetes and related bacteria

Order: Actinomycetales
    Family: Actinomycetaceae
        Genus: *Actinomyces (A. bovis)*
                Bacterionema (B. matruchoti)
                            Arachnia (A. propionica)
                            Bifidobacterium (B. bifidum)

Rothia (R. dentocariosa)

Family: Actinoplanaceae
Genus: *Actinoplanes (A. philippinensis)
Ampullariella (A. regularis)
Kitasatoa (K. purpurea)
Planobispora (P. longispora)
Spirillospora (S. albida)
Amorphosporangium (A. auranticolor)
Dactylosporangium (D. aurantiacum)
Pilimelia (P. terevasa)
Planomonospora (P. parontospora)
Streptoalloteichus (S. hindustanus)

Family: Streptosporangiaceae
Genus: *Steptosporangium (S. roseum)

Family: Cellulomonadaceae
Genus: *Cellulomonas (C. flavigena)
Oerskovia (O. turbata)
Jonesia (J. denitrificans)
Promicromonospora (P. citrea)

Family: Dermatophilaceae
Genus: *Dermatophilus (D. congolensis)
Related genus: Conglomeromonas (C. largomobilis)
Geodermatophilus (G. obscurus)

Family: Frankiaceae
Genus: *Frankia (F. alni)

Family: Micromonosporaceae
Genus: *Micromonospora (M. chalcea)
Micropolyspora (M. brevicatena)
Thermomonospora (T. curvata)
Microbispora (M. rosea)
Thermoactinomyces (T. vulgaris)

Related genus: Actinobispora (A. yunnanensis)
Faenia (F. rectivirgula)

Family: Nocardiaceae
Genus: *Nocardia (N. asteroides)
Saccharopolyspora (S. hirsuta)
Actinopolyspora (A. halophila)

Related genus: Amycolata (A. autotrophica)
Kibdelosporangium (K. aridus)
Amycolatopsis (A. orientalis)
Pseudoamycolata (P. halophobica)

Family: Pseudonocardiaceae
Genus: *Pseudonocardia (P. thermophila)

Family: Streptomycetaceae
Genus: *Streptomyces (S. albus)
Actinosporangium (A. violaceum)
Sporochthya (S. polymorpha)
Actinopycnidium (A. caeruleum)
Microellobosporia (M. cinerea)
Streptoverticillium (S. baldaccii)

Related genus: Chainia (C. antibiotica)
                Intrasporangium (I. calvum)
Family: Nocardioidaceae
Genus: *Nocardioides (N. albus)
Related genus: Terrabacter (T. tumescens)
Genera of uncertain affiliation:
                Actinokineospora (A. riparia)
                Glycomyces (G. harbinensis)
                Elytrosporangium (E. brasiliensis)

Order: Mycobacteriales
Family: Mycobacteriaceae
Genus: *Mycobacterium (M. tuberculosis)
                Catellatospora (C. citrea)
                Sphaerobacter (S. thermophilus)

Order: Caryophanales
Family: Caryophanaceae
Genus: *Caryophanon (C. latum)
Genera of uncertain affiliation:
                Actinomadura (A. madurae)
                Coriobacterium (C. glomerans)
                Kitasatosporia (K. setalba)
                Nocardiopsis (N. dassonvillei)
                Saccharomonospora (S. viridis)
                Actinosynnema (A. mirum)
                Kineosporia (K. aurantiaca)
                Microtetraspora (M. glauca)
                Renibacterium (R. salmoninarum)
                Saccharothrix (S. australiensis)

Coryneform bacteria
Family: Brevibacteriaceae
Genus: *Brevibacterium (B. linens)
Family: Corynebacteriaceae
Genus: *Corynebacterium (C. diphtheriae)
Family: Propionibacteriaceae
Genus: *Propionibacterium (P. freundenreichii)
                Acetogenium (A. kivui)
                Acetobacterium (A. woodii)

Order: Eubacteriales
Tribe: Eubacterieae

Genus: *Eubacterium (E. foedans)
Genera of uncertain affiliation:

Aeromicrobium (A. erythreum) Arcanobacterium (A. haemolyticum)
Arthrobacter (A. globiformis) Aureobacterium (A. liquefaciens)
Brachybacterium (B. faecium) Caseobacter (C. polymorphus)
Clavibacter (C. michiganense)
Curtobacterium (C. citreum) Dermabacter (D. hominus)
Exiguobacterium (E. aurantiacum) Kurthia (K. zopfii)
Microbacterium (M. lacticum) Pimelobacter (P. simplex)
Rarobacter (R. faecitabidus) Rubrobacter (R. radiotolerans)
Tsukamurella (T. paurometabolum)

The spirochaetes

Order:
Family: Spirochaetales
  Leptospiraceae
Genus: *Leptospira (L. interrogans) Leptonema (L. illini)
Family: Spirochaetaceae
Genus: *Spirochaeta (S. plicatilis) Borrelia (B. anserina)
  Brachyspira (B. aalborgi) Clevelandina
  (C. reticulitermitidis)
  Crispispira (C. pectinis) Diplocalyx (D. calotermitidis)
  Hollandina (H. pterotermitidis) Pilloteria (P. calotermitidis)
Family; Treponemataceae
Genus: *Treponema (T. pallidum)
Related genus: Serpulina(S. hyodysenteriae)

The rickettsias

Order:
Family: Rickettsiales
  Anaplasmataceae
Genus: *Anaplasma (A. marginale) Aegyptianella (A. pullorum)
  Eperythrozoon (E. coccoides) Haemobartonella (H. muris)

Family:	Bartonellaceae	
Genus:	*Bartonella (B. bacilliformis)	
Family:	Ehrlichiaceae	
Genus:	*Ehrlichia (E. canis)	Grahamella (G. talpae)
	Neorickettsia (N. helminthoeca)	Cowdria (C. runinantium)
Family:	Rickettsiaceae	
Genus:	*Rickettsia (R. prowazekii)	Coxiella (C. burnetii)
	Piscirickettsia (P. salmonis)	Rochalimaea (R. quinana)
Tribe:	Wolbachieae	
Genus:	*Wolbachia (W. pipientis)	Blattabacterium (B. cuenoti)
	Rickettsiella (R. popilliae)	Symbiotes (S. lectularius)

Order: Chlamydiales
Family: Chlamydiaceae
Genus: *Chlamydia (C. trachomatis)

The mycoplasmas

Order: Acholeplasmatales
Family: Acholeplasmataceae
Genus: *Acholeplasma (A. laidlawii)

Order: Mycoplasmatales

Family:	Mycoplasmataceae	
Genus:	*Mycoplasma (M. mycoides)	Ureoplasma (U. urealyticum)
Family:	Spiroplasmataceae	
Genus:	*Spiroplasma (S. citri)	
Genera of uncertain affiliation:		
	Anaeroplasma (A. abactoclasticum)	Thermoplasma (A. acidophilum)

Endosymbionts

Genus:

Buchnera (B. aphidicola)	Caedibacter (C. taeniospiralis)
	Holospora (H. undulata)
Polynucleobacter (P. necessarius)	Pseudocaedibacter (P. conjugatus)
Sarcobium (S. lyticum)	Tectibacter (T. vulgaris)

Miscellaneous genera

Ancalochloris (A. perfilievii)	Bactoderma (B. alba)
Chloronema (C. giganteum)	Excellospora (E. viridilutea)
Sphingomonas (S. paucimobilis)	Stibiobacter (S. senarmontii)

Appendix B: Classification of the bacteria – the phylogenetic approach

This scheme has been based on phylogenetic relationships, as discussed by Stackebrandt in, *The Prokaryotes*, 2nd edition, (A. Balows, H.G. Trüper, M. Dworkin, W. Harder and K.H. Schleifer, Eds., pp. 19–46. Springer-Verlag, New York) and expanded to include the likely location of groups from the 'conventional' data in *Bergey's Manual of Systematic Bacteriology*, Vol. 1 to 4, and the *Approved Lists of Bacterial Names* (1980) and their supplements.

Gram-positive Eubacteria

Actinomycete branch

Family:	Actinomaduraceae	
Genus:	Actinomadura	
	Thermomonospora	
Family:	Actinoplanaceae	
Genus:	Actinoplanes	
	Ampullariella	
	Kitasatoa	
	Spirillospora	
	Micromonospora	
	Actinomyces	
Related genera:	Actinobispora	
	Aureobacterium	Amorphosporangium
	Brevibacterium	Dactylosporangium
		Pilimelia
		Streptoalloteichus
		Micropolyspora
		Arthrobacter
		Bifidobacterium
Family:	Cellulomonadaceae	
Genus:	Cellulomonas	
	Oerskovia	Jonesia
		Promicromonospora

Related genera: Aeromicrobium   Arcanobacterium
    Brachybacterium   Caseobacter
    Clavibacter   Conglomeromonas
    Corynebacterium   Curtobacterium
    Dermabacter   Dermatophilus
    Exiguobacterium   Pimelobacter
    Rarobacter   Rubrobacter

Family: Frankiaceae
Genus: Frankia
    Blastococcus   Geodermatophilus

Maduromycetes

Genus: Microbispora   Planobispora
Related genera: Planomonospora
    Microbacterium   Micrococcus
    Mobiluncus   Mycobacterium
    Paracoccus

Nocardioforms

Genus: Nocardia
Related genera: Nocardioides   Tsukamurella
    Propionibacterium   Nocardiopsis
    Pseudonocardiaceae   Pseudoamycolata
Family: Pseudonocardia   Actinopolyspora
Genus: Amycolata   Amycolatopsis
    Faenia   Kibdelosporangium
    Saccharomonospora   Saccharopolyspora
Related genera: Renibacterium   Rothia
    Saccharothrix   Sphaerobacter
    Stomatococcus

Family: Streptomycetaceae
Genus: Streptomyces   Actinopycnidium
    Actinosporangium   Microellobosporia
    Sporochthya   Streptoverticillium

Related genera: Chainia
                 Intrasporangium
                 Elytrosporangium
                 Terrabacter

Other genera containing 'actinomycete':
       Actinokineospora   Actinosynnema
       Arachnia   Bacterionema
       Catellatospora   Caryophanon
       Coriobacterium   Glycomyces
       Kineosporia   Kitasatosporia
       Microtetraspora   Streptosporangium

Clostridium–Bacillus branch

  Genus: Acetobacterium   Acetogenium
               Amphibacillus   Bacillus
               Brochothrix   Enterococcus
               Clostridium   Desulfotomaculum
               Eubacterium   Filibacter
               Gemella

  Family: Haloanaerobiaceae
  Genus: Haloanaerobium   Halobacteroides
               Haloincola

  Related genera: Alloiococcus   Carnobacterium
                     Kurthia   Lactobacillus
                     Lactococcus   Leuconostoc
                     Listeria

Order:
  Genus: Mycoplasmatales   Acholeplasma
               Mycoplasma   Erysipelothrix
               Anaeroplasma   Spiroplasma
               Ureoplasma

  Related genera: Aerococcus   Coprococcus
                     Hyperthermus   Marinococcus
                     Melisococcus   Pediococcus
                     Peptococcus   Peptostreptococcus

Planococcus
Saccharococcus
Sarcina
Sporolactobacillus
Staphylococcus
Syntrophobacter
Thermoactinomyces
Thermoanaerobacter
Trichococcus

Ruminococcus
Salinicoccus
Sporohalobacter
Sporosarcina
Streptococcus
Syntrophospora
Thermoanaerobium
Thermobacteroides
Vagococcus

Gram-positive Eubacteria with Gram-negative cell wall
Genus:
Acidaminococcus
Gemminger
Heliothrix
Llyobacter
Megasphaera
Pectinatus
Selenomonas
Zymophilus

Butyrivibrio
Heliobacterium
Lachnospira
Malonomonas
Oxalobacter
Roseburia
Veillonella

Proteobacteria branch
Alpha subclass
Genus:
Acetobacter
Ancalomicrobium
Azospirillum
Brucella
Erythrobacter
Gemmobacter
Hyphomicrobium
Methylobacterium
Nitrobacter
Pedomicrobium
Prosthecomicrobium

Acidomonas
Aquaspirillum
Beijerinckia
Caulobacter
Filomicrobium
Gluconobacter
Hyphomonas
Methylomonas
Paracoccus
Phenylobacterium
Pseudomonas

Rhodobacter
Rhodopila
Rhodospirillum

Rhodomicrobium
Rhodopseudomonas

Other genera containing ammonia or nitrite oxidizing bacteria:
Nitrococcus
Nitrospira

Nitrospina

Other genera containing budding and appendaged bacteria:
Angulomicrobium
Dichotomicrobium
Gallionella
Labrys
Pasteuria

Asticcacaulis
Ensifer
Hirschia
Nevskia
Prosthecobacter

Other genera from the Order Spirillales:
Halovibrio
Magnetospirillum
Vampirovibrio

Herbaspirillum
Prolinoborus

Other genera containing phototrophs:
Lamprobacter
Prochloron
Rhodoferax
Rubrivivax

Pelodictyon
Prochlorothrix
Roseobacter

Family: Rhizobiaceae
Genus: Rhizobium
             Azorhizobium
             Bradyrhizobium
             Ochrobacterium

Agrobacterium
Blastobacter
Phyllobacterium
Sinorhizobium

Related genera:

Order: Rickettsiales
Genus: Rickettsia
Other genera containing 'rickettsias':
             Anaplasma
             Bartonella
             Cowdria

Rochalimaea

Aegyptianella
Blattabacterium
Coxiella

Other genera: *Ehrlichia* *Eperythrozoon*
Grahamella *Haemobartonella*
Neorickettsia *Piscirickettsia*
Rickettsiella *Symbiotes*
Wolbachia
Seliberia *Stella*
Thiobacillus *Xanthobacter*
Zymomonas

Beta subclass
Genus: *Achromobacter*
Family: *Alcaligenaceae*
Genus: *Alcaligenes*
Other genera: *Acidovorax* *Bordetella*
Comamonas *Chromobacterium*
Hydrogenophaga *Derxia*
Leptothrix *Janthinobacterium*
Variovorax

Family: *Neisseriaceae*
Genus: *Neisseria* *Alysiella*
Eikenella *Kingella*
Other genera: *Mesophilobacter* *Morococcus*
Nitrosococcus *Nitrosolobus*
Nitrosomonas *Nitrosospira*
Oligella *Rhodocyclus*
Sphaerotilus *Spirillum*
Taylorella

Other genera containing sheathed bacteria:
Crenotrix

Gamma subclass
Genus: *Acinetobacter* *Aeromonas*
Alteromonas *Azomonotrichon*
Azorhizophilus *Azotobacter*

Genus: Azomonas
Cardiobacterium
Suttonella
Branhamella
Dichelobacter

Family: Chromatiaceae
Genus: Chromatium
Lamprobacter
Thiocapsa
Thiodictyon
Thiospirillum
Amoebobacter
Lamprocystis
Thiocystis
Thiopedia

Family: Ectothiorhodospiraceae
Genus: Ectothiorhodospira

Family: Enterobacteriaceae
Genus: Escherichia
Budvicia
Cedecea
Edwardsiella
Erwinia
Hafnia
Kluyvera
Leclercia
Levinea
Morganella
Pantoea
Pragia
Providencia
Salmonella
Shigella
Trabulsiella
Yersinia
Arsenophonus
Buttiauxella
Citrobacter
Enterobacter
Ewingella
Klebsiella
Koserella
Leminorella
Moellerella
Obesumbacterium
Pectobacterium
Proteus
Rahnella
Serratia
Tatumella
Xenorhabdus
Yokenella

Related genus: Saccharobacter
Other genera: Frateuria
Zoogloea

Family: Halomonadaceae
Genus: Halomonas
Deleya

Related genus:	*Marinobacter*	*Fluoribacter*
Family:	*Legionellaceae*	
Genus:	*Legionella*	
	Tatlockia	
Other genera:	*Leucothrix*	*Lysobacter*
	Marinomonas	*Methylococcus*
	Moraxella	*Oceanospirillum*
	Psychobacter	*Volcaniella*
Family:	*Pasteurellaceae*	
Genus:	*Pasteurella*	*Actinobacillus*
	Haemophilus	
Related genus:	*Gardnerella*	
Other genera:	*Plesiomonas*	*Ruminobacter*
	Serpens	*Thiothrix*
Family:	*Vibrionaceae*	
Genus:	*Vibrio*	*Allomonas*
	Enhydrobacter	*Listonella*
	Photobacterium	*Shewanella*
		Micavibrio
		Xanthomonas
		Xylophilus
Related genera:	*Colwellia*	
Other genera:	*Vitreoscilla*	
	Xylella	

Delta subclass

Genus:	*Bdellovibrio*	
Family:	*Myxococcaceae*	
Genus:	*Myxococcus*	*Angiococcus*
	Chondromyces	*Cystobacter*
	Nannocystis	*Stigmatella*
Other genera containing gliding bacteria:	*Achromatium*	*Archangium*
	Chitinophaga	*Melittangium*
	Microscilla	*Polyangium*
	Simonsiella	*Sphingobacterium*

Other genera:

Thermonema
Desulfobacter
Desulfobulbus
Desulfohalobium
Desulfomonile
Desulfosarcina
Desulfuromonas

Toxothrix
Desulfobacterium
Desulfococcus
Desulfomonas
Desulfonema
Desulfovibrio
Pelobacter

Campylobacter branch
Genus:

Arcobacter
Helicobacter
Wolinella

Campylobacter
Thiovulum

Spirochaetes branch
Genus:

Borrelia
Spirochaeta

Leptonema
Treponema

Other genera containing 'spirochaetes':

Braychyspira
Crispispira
Hollandina
Pilloteria

Clevelandina
Diplocalyx
Leptospira
Serpulina

Chlorobiaceae branch
Genus:

Chlorobium
Flectobacillus
Spirosoma

Chloroherpeton
Prosthecochloris

Related genera:

Ancylobacter
Cyclobacterium
Pelosigma

Brachyarcus
Meniscus
Runella

Bacteroides/Cytophaga branch
Genus:

Acetivibrio
Acetomicrobium

Acetoanaerobium
Acidaminobacter

Anaerobiospirillum   Anaerovibrio
Bacteroides   Centipeda
Cytophaga   Capnocytophaga
Flavobacterium   Flexibacter
Flexithrix   Fusobacterium
Haliscomenobacter   Leptotrichia
Porphyromonas   Propionigenium
Propionispira   Rikenella
Saprospira   Sporocytophaga
Succinivibrio   Succinomonas
Syntrophomonas   Syntrophus

Planctomyces branch
  Genus: *Gemmata*   *Planctomyces*
                  Pirellula

Chlamydia branch
  Genus: *Chlamydia*

Deinococcus branch
  Genus: *Deinococcus*   *Thermus*
  Related genus: *Deinobacter*

Chloroflexus branch
  Genus: *Chloroflexus*   *Herpetosiphon*
                  Thermomicrobium

Verrucomicrobium branch
  Genus: *Verrucomicrobium*

Thermotoga branch
  Genus: *Thermotoga*   *Ferroidobacterium*
                  Thermosiphon

Related genera: *Aquifex* *Pyrobaculum*

Archaebacteria

Methanogens and halophiles branch
Genus:
Haloarcula	*Halobacterium*
Halococcus	*Haloferax*
Methanobacterium	*Methanobrevibacter*
Methanococcoides	*Methanococcus*
Methanocorpusculum	*Methanohalobium*
Methanogenium	*Methanolobus*
Methanomicrobium	*Methanoplanus*
Methanosarcina	*Methanospirillum*
Methanothermus	*Methanothrix*
Natronobacterium	*Natronococcus*
Thermoplasma	

Related genera:
Halomethanococcus	*Methanoculleus*
Methanohalophilus	*Methanolacinia*
Methanopyrus	*Methanosphaera*
Syntrophococcus	

Thermococcus branch
Genus: *Thermococcus* *Pyrococcus*

Thermoacidophiles branch
Genus:
Desulfurococcus	*Desulfurolobus*
Pyrodictium	*Sulfolobus*
Stygiolobus	*Thermoproteus*
Thermophilum	

Other sulphur bacteria:
Acidianus	*Archaeoglobus*
Macromonas	*Metallosphaera*
Thermothrix	*Thiomicrospira*
Thiosphaera	*Thiospira*

Hydrogen bacteria
Genus: *Acetonema*   *Hydrogenobacter*
           Calderobacterium   *Hydrogenovibrio*

Endosymbionts
Genus: *Buchnera*   *Caedibacter*
           Holospora   *Polynucleobacter*
           Pseudocaedibacter   *Sarcobium*
           Tectibacter

Miscellaneous groups

Acidophilium	*Acetofilamentum*
Acetohalobium	*Acetothermus*
Acidothermus	*Agitococcus*
Agromonas	*Aminobacter*
Ancalochloris	*Bactoderma*
Bilophila	*Calymmatobacterium*
Capsularis	*Carboxydothermus*
Chloronema	*Chromohalobacter*
Chryseomonas	*Cupriavidus*
Excellospora	*Falcivibrio*
Fibrobacter	*Flavimonas*
Formivibrio	*Francisella*
Iodobacter	*Lampropedia*
Megamonas	*Methylobacillus*
Methylophaga	*Methylophilus*
Methylovorus	*Mitsuokella*
Prevotella	*Protomonas*
Rhizobacter	*Rhizomonas*
Rugamonas	*Sphingomonas*
Stibiobacter	*Streptobacillus*
Sulfobacillus	*Thermoleophilum*
Weeksella	

Family: *Oscillospiraceae*

Index